普通高等教育"十三五"规划教材

普通化学实验英文教程
General Chemistry Experiments

耿旺昌　颜　静　陈　芳　王　欣　编著

电子工业出版社
Publishing House of Electronics Industry
北京·BEIJING

内 容 简 介

本书根据高等院校工科专业学生对化学实验学习的需求及特点编写而成。本实验教材与大学普通化学课程相对应。教材分为 4 部分：第 1 部分简要介绍实验室安全、基本实验操作及数据处理；第 2 部分介绍 10 个不同的实验，这些实验分别与普通化学教学中的化学热力学、化学动力学、溶液、电化学、配位化合物等理论内容相对应；第 3 部分是附录，包含实验中用到的一些数据；第 4 部分是化学实验报告。

本书既可作为高等院校各工科专业的基础化学实验教学用书，又可供自学者、工程技术人员参考使用。

未经许可，不得以任何方式复制或抄袭本书之部分或全部内容。
版权所有，侵权必究。

图书在版编目（CIP）数据

普通化学实验英文教程 / 耿旺昌等编著. —北京：电子工业出版社，2019.3
ISBN 978-7-121-35998-9

I. ①普⋯ II. ①耿⋯ III. ①普通化学—化学实验—高等学校—教材—英文 IV. ①O6-3

中国版本图书馆 CIP 数据核字（2019）第 012171 号

策划编辑：孟　宇
责任编辑：章海涛
印　　刷：北京虎彩文化传播有限公司
装　　订：北京虎彩文化传播有限公司
出版发行：电子工业出版社
　　　　　北京市海淀区万寿路 173 信箱　邮编：100036
开　　本：787×1092　1/16　印张：8.25　字数：170 千字
版　　次：2019 年 3 月第 1 版
印　　次：2019 年 6 月第 2 次印刷
定　　价：30.00 元

凡所购买电子工业出版社图书有缺损问题，请向购买书店调换。若书店售缺，请与本社发行部联系，联系及邮购电话：(010) 88254888，88258888。
质量投诉请发邮件至 zlts@phei.com.cn，盗版侵权举报请发邮件至 dbqq@phei.com.cn。
本书咨询联系方式：mengyu@phei.com.cn。

Preface

General chemistry and general chemistry experiments are important curriculum in an engineering university for students majoring in engineering and chemistry related. The general chemistry experiments course aims to further verify the fundamental principles of chemistry showing in general chemistry class, training professional skills by practice, provides a good link between theory and practice in general chemistry, and cultivate scientific approach in seeking the truth from facts. Through this class, students can finally improve their ability of practice, observe and analysis when face to problems in further study and daily lives.

This book was designed to help students majoring in engineering or biology to find their special interests in chemistry experiments as non-professional chemists. This book includes 10 experiments. Dr. Wangchang Geng provided the writing outlines and requirements. Experiments 1~3 were edited by Dr. Fang Chen. Experiments 4~6 were edited by Dr. Wangchang Geng. Experiments 7~10 were edited by Dr. Jing Yan. Dr. Xin Wang was responsible for the correction of this textbook. Authors would like to thank these teachers from the Teaching and Research Group of General Chemistry in NPU for their generous help with the preparation of this book.

Your comments or feedback about this book are most welcome.

Wangchang Geng   Jing Yan   Fang Chen   Xin Wang

Contents

Part 1　General Principles in Chemical Laboratory ·················· 1
　　Chapter 1　Chemical Laboratory Safety Rules ··················· 1
　　Chapter 2　General Operations ··························· 6
　　Chapter 3　Data and Error Analysis ························ 12

Part 2　Experiments ··· 15
　　Experiment 1　Determination of the Ideal Gas Constant ············ 15
　　Experiment 2　Determination of Enthalpy of Chemical Reaction— Zn and $CuSO_4$ Reaction ································ 23
　　Experiment 3　Chemical Reaction Rate ······················· 28
　　Experiment 4　Electrochemistry ···························· 36
　　Experiment 5　Inorganic Compound ························ 44
　　Experiment 6　Determination of Dissociation Constant of Bromothymol Blue by Spectrophotometric Method ························ 51
　　Experiment 7　Synthesis of Potassium Trioxalatoferrate (III) ········· 63
　　Experiment 8　Determination of Calcium and Magnesium Ion in Water by Complexometric Titrations with EDTA ··················· 66
　　Experiment 9　Determination of Mo (VI) Content by Visual Catalytic Kinetics Method ··························· 72
　　Experiment 10　Separation of Coordination Compounds by Ion Exchange Resin ··················· 78

Part 3　Appendices ··· 83
　　Appendix 1　Specification and Selection of Chemical Reagent ········ 83
　　Appendix 2　Density and Concentration of Commonly Used Acid or Base Solutions (15 ℃) ··················· 84

Appendix 3	Color of Common Ions	85
Appendix 4	International Relative Atomic Mass	86
Appendix 5	Standard Electrode Potential	88
Appendix 6	Saturated Vapor Pressure of Water at Different Temperature	92
Appendix 7	Stability Constant of Some Coordinated Ions	94
Appendix 8	Boiling Point of Water at Different Pressure	95
Appendix 9	Density of Water at Different Temperature	96
Appendix 10	Relationship Between pH and Temperature of Buffer Solutions	97
Appendix 11	Dissociation Constant of Some Common Weak Electrolytes (298.15 K)	98
Appendix 12	Solubility Product of Some Common Substances	99
Appendix 13	Symbols, Values and SI Units of Some Constants	100

Part 4　Experimental Reports ··· 101

Experimental Report 1 ··· 101
Experimental Report 2 ··· 103
Experimental Report 3 ··· 106
Experimental Report 4 ··· 110
Experimental Report 5 ··· 112
Experimental Report 6 ··· 114
Experimental Report 7 ··· 117
Experimental Report 8 ··· 119
Experimental Report 9 ··· 121
Experimental Report 10 ··· 124

Part 1　General Principles in Chemical Laboratory

Chapter 1　Chemical Laboratory Safety Rules

Accidents in a chemical laboratory usually are resulted from improper judgment of the victim or one of his/her neighbors. Learn and observe the safety and laboratory rules listed below before starting to do chemical experiments.

Precautions

(1) **Maintain a wholesome, businesslike attitude in the laloratory.** Horseplay and other acts of carelessness are prohibited.

(2) **Never work in the laloratory without the instructor present.** This includes setting up equipment.

(3) **Wear suitable clothing.** Wear clothing that will protect you against spilled chemicals or flaming liquids. Hard-soled, covered footwear must be worn in the laboratory at all times——**no sandals allowed**.

(4) **Eating, drinking, and smoking are strictly prohibited in the laboratory at all times** because of the possibility of chemicals getting into the mouth or lungs through contamination. The chief hazard with smoking is fire or explosion.

(5) **Know the ways to put out a fire.**

a) If it is open fire, such as a large chemical spill on a laloratory bench, the correct extinguisher should be used as follows:
- Pull the pin.
- Point the extinguisher (if dry) or hose (if CO_2) at the base of the fire.
- Squeeze the handle while moving the extinguisher back and forth.

NOTE: Be careful not to spread the fire by getting the nozzle of the extinguisher too close——the material being emitted is under pressure.

b) If it is a small contained fire, such as in a flask or beaker, cover the container with a piece of ceramic, cutting off the supply of oxygen to the fire and thus putting it out.

(6) Wash chemicals from skin.

a) If you get a chemical burn from a caustic material, i.e. acid or base, immediately wash the burned area with large amount of water. Ask another student to inform the laloratory instructor.

b) Wash your hands and face quickly and thoroughly whenever they come into contact with chemicals.

c) Always wash your hands, before leaving the laloratory since toxic chemicals may be transferred to your mouth at a later time.

d) Chemicals spilled over a large part of the body require immediate action. Remove all contaminated clothing and use the safety shower, flooding the burned area. Do not use salves, creams, lotions, etc. Get medical attention.

Preparation before the Class

A detailed preparation is highly requested before each single experiment. The experimental instructor may list the order of all the experiments before the first class.

Rules during the Experiment

(1) Keep your workspace orderly.

a) Place tall items, such as graduated cylinders, toward the back of the workbench so they will not be overturned by reaching over them.

b) Clean up all chemical spills, scraps of paper, and glasswares immediately.

c) Keep drawers closed while working and the aisles free of any obstructions.

d) Never place coats, books, and other belongings on the laboratory bench where they will interfere with the experiment and are likely to be damaged.

(2) Always read the label first. Identify the chemicals before use. Read the label carefully, *read it twice*, before taking anything from a bottle. Many chemicals have similar names, such as sodium sulfate and sodium sulfite. Using the wrong reagent can spoil an experiment or can cause a serious accident.

(3) Assume that a particular reagent is hazardous unless you know for sure it is not.

a) Never taste a chemical unless specifically instruction to do so.

b) If you are instructed to smell a chemical, point the vessel away from your face and carefully fan the vapors toward your face with your hand and sniff gently.

c) Material Safety Data Sheets are available.

(4) Never point a test tube toward your neighbors or yourself when:

a) Heating a test tube over a burner.

b) Carrying out a reaction in a test tube.

(5) Dilute concentrated acids and bases by pouring the reagent into water (room temperature or lower) while stirring constantly. Never pour water into concentrated acids; the heat of solution will cause the water to boil and the acid to splatter. To help you remember——*"Do as you oughter, pour acid into water."*

Be careful with flames. A lighted gas burner can be a major fire hazard.

a) General Precautions:
- The burner should be burning only for the period of time in which it is actually utilized.
- Before lighting your burner carefully, position it on the desk away from flammable materials, overhanging reagent shelves, flammable

reagents such as acetone, toluene, and alcohol on neighboring desks.
- Be careful not to extend your arm over a burner while reaching for something.

b) Personal Precautions:
- Keep long hair tied back so that it cannot fall forward into a flame.
- Keep beards away from flames.

(6) Assemble safe apparatus. Always assemble an apparatus as outlined in your instructions. Makeshift equipment and poor apparatus assemblies are the first step to an accident.

(7) Avoid rubbing your eyes unless you know your hands are clean.

(8) Do not put hot objects on the desktops. Place hot objects on a wire gauze or ceramic pad.

(9) Never throw lighted matches into a sink. They may ignite a discarded flammable liquid.

(10) Perform only authorized experiments. Unless authorized to do so by the experimental instructor, a student will be subject to immediate and permanent expulsion from the laloratory if:

a) Attempting to conduct unauthorized experiments.

b) Attempting variations of the experiment in the laloratory manual.

Performing unauthorized experiments are dangerous. Students lack the experience to recognize whether or not the chemicals and techniques are safe.

(11) Clean up your workspace at the end of each laboratory period.

a) Wash and wipe off your desktop.

b) Be sure gas and water is turned off.

c) Return all special equipment to the stockroom.

(12) Avoid using excessive amounts of reagent.

a) Never use more than called for in the experiment.

b) Do not return any excess chemical to the reagent bottle; share it with another student or dispose of it according the in the instructions listed in (13).

c) If you are uncertain how to dispose of an excess of a specific chemical, consult your experimental instructor.

(13) Discard waste chemicals as follows:

Waste Chemical's Proper Disposal.

a) Non-flammable water-soluble liquids discard into liquid wastes bottle.

b) Chemical solids, contaminated paper, and contaminated broken glassware, solid wastes bottle.

c) Discard paper products into trash can.

d) Discard organic solvents into organic waste bottle (Do not put acids in the organic waste bottle).

e) Discard glass tubing waste or broken glass wares into broken glass wooden box.

(14) Always add a reagent slowly——never "dump" in. There're two reasons:

a) Some reactions give off a lot of heat, and unless adding slowly, can become too vigorous and out of control.

b) If you make a mistake and choose the wrong chemical, adding slowly decreases the possibility of causing a serious accident.

(15) Treat chemical spills as follows:

a) Alert your neighbors and your experimental instructor.

b) Clean up the spill as directed by your experimental instructor.

(16) Never fill a vessel more than about 70% capacity if you plan to heat it, unless specifically told to do so.

(17) Be aware of your neighbors' activities; you may be a victim of their mistakes. If you observe improper techniques or unsafe practices:

a) Advise your neighbors.

b) Advise your experimental instructor if necessary.

(18) Observe all specific precautions and modifications mentioned in each experiment.

(19) Do not remove any chemicals from the laboratory.

(20) For reasons of safety, you may not be allowed to enter the laboratory if you are late.

Chapter 2　General Operations

1. Washing Glassware

In order to obtain a precise results, cleanliness of glassware is highly required.

(1) Chemical stains are often cleaned up with acids, alkalis, or organic solvents.

(2) For example, the stain of the azo dyes (Methyl Red and Methyl Orange) prepared in "Organic Chemistry Experiment" can be easily dissolved in an alkaline solution.

(3) Do not scrub the inside of volumetric glassware such as burettes, volumetric pipettes, measuring flasks, and conical measures with a brush. This could cause a volumetric disorder. Volumetric glassware should be rinsed repeatedly with tap water.

(4) For the same reason, volumetric glassware should not be dried by heating.

(5) Wash off chemical stains with a brush and sodium hydrogen carbonate.

(6) Dissolve stains of organic substances with a small amount of ethanol. Pour the washing waste of ethanol into the specific waste container.

2. Liquid Wastes

Laboratory liquid wastes are classified into organic and inorganic wastes. Pour liquid wastes into the designated liquid wastes container according to textbooks and experimental instructor. Rinse the glassware with the minimum amount of water, and pour into the liquid wastes container.

3. Volumetric pipette

The volume of a volumetric pipette (Fig. 1-1) is the volume of a liquid going out completely from the pipette which is filled up to the marked line. In order to avoid volume disorder, volumetric laboratory glassware such as volumetric pipettes should not be dried by heating. The wet glassware are rinsed with the liquid to be used just before use.

Fig. 1-1 Volumetric pipette

Volumetric Pipette Operation Procedures

(1) Insert the tip of the pipette into the liquid deeply, and draw up it 1-2 cm above the marked line.

(2) After pulling out the tip of the pipette above the surface, drop the inside liquid slowly to adjust the meniscus precisely to the marked line.

(3) Next, put the tip of the pipette on the inside surface of the container to remove a remaining droplet.

(4) Keep the pipette perpendicularly and allow the liquid to fall naturally into the container.

(5) After almost liquid has been drained, hold the bulge of the pipette in your palm during which all the valves of the filler are closed. Then, the inside air expands and all of the liquid falls.

4. Operation of Volumetric Flasks

A volumetric flask is employed when a solution of a definite concentration is prepared. The volume of a volumetric flask is the volume of a liquid in the flask which is filled up to the marked line. When a standard solution is prepared from a solid sample, it is not dissolved in a volumetric flask but in another container. Then, the solution is transferred into the volumetric flask. Because the volume may change at mixing of a sample and a solvent, the sample should not be diluted directly to the marked line without stopping. After the final dilution, the solution is mixed thoroughly, by inverting the flask and shaking. The mixing is not enough when it is shaken without inverting. The way of determine the finishing point in volumetric flask is shown in Fig. 1-2.

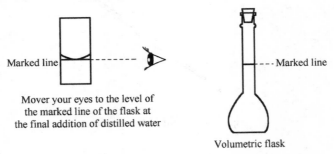

Fig. 1-2　The way of determine the finishing point in volumetric flask

Volumetric Flask Operation Procedures

(1) To make up a standard solution, add first a less amount of distilled water than required to the solid sample weighted precisely.

(2) Next, stir it to be dissolved completely using a stirring stick.

(3) Transfer the solution into the volumetric flask with a stirring stick or a funnel.

(4) In order to transfer the remaining solution in the container and on the stirring stick, rinse them some times with distilled water.

(5) Shake the flask gently to make the solution homogeneous.

(6) Repeat the operation of adding a small amount of distilled water and shaking until the height of the solution reaches just below the marked line.

(7) Move your eyes to the level of the marked line of the flask at the final addition of distilled water. Adjust the meniscus of the solution to the line by adding distilled water dropwise with a transfer pipette.

(8) After the final dilution, mix the solution thoroughly by inverting the flask and shaking with holding tightly the stopper.

Tuning procedures as using volumetric flask is shown in Fig. 1-3.

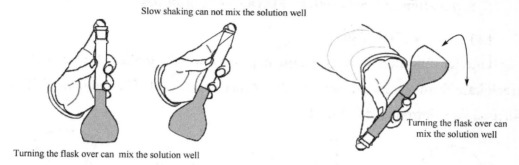

Fig. 1-3 Tuning procedures as using volumetric flask

5. Burette

Burettes are graduated cylinders that measure the volume of the drained liquid. Burettes should not be dried by heating, and thus you may use immediately after rinsing with small amounts of the solution to be measured.

Two approaches of burette operations are shown in Fig. 1-4.

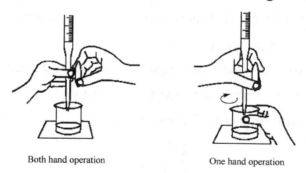

Fig. 1-4 Two approaches of burette operations

Burette Operation Procedures

(1) To transfer the titration solution into a burette, close the stopcock at the bottom and use a funnel. Do not forget to remove the funnel after transferring the solution.

(2) Drain a small amount of the solution to expel air bubbles at the tip of the burette. You do not need to adjust the liquid surface to 0.00 mL.

(3) Move your eyes to the level of the liquid surface and read the value of the bottom of the meniscus with 1/10 of the smallest scale marked on the burette.

6. Separation of Precipitates - Filtration Techniques

(A) Gravity Filtration

Gravity filtration is a common approach which applied for separating a precipitate from a solution using a filter paper and funnel. The method of folding filter paper is introduced in Fig. 1-5.

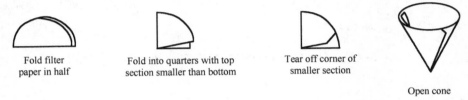

Fold filter paper in half　　Fold into quarters with top section smaller than bottom　　Tear off corner of smaller section　　Open cone

Fig. 1-5　The method of folding filter paper

Gravity Filtration Operation Procedures (Fig. 1-6)

(1) Choose a filter paper. Fold the filter paper into four and fit it into the funnel.

(2) When the filter paper does not fit well into the funnel, slightly adjust the folding in order to fit it perfectly.

(3) In order to make the filter paper fit perfectly, cut the edge of the over lapping filter paper in contact with the funnel obliquely. Wet the paper with the solvent and press with fingers to fit the paper in.

Fig. 1-6　Gravity filtration operation procedures

(4) Place a container in a manner that the leg of the funnel touches its inside wall.

(B) Vacuum Filtration

Vacuum filtration rather than gravity filtration is suitable for the separation of a precipitate consisting of light and fine particles. The vacuum filtration apparatus could be illustrated as shown in Fig. 1-7.

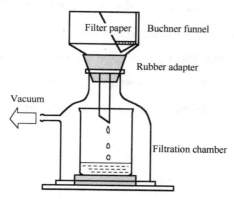

Fig. 1-7 The vacuum filtration apparatus

Vacuum Filtration Operation Procedures

(1) Connect a filtration chamber and an aspirator with thick wall rubber tubing through a three-way stopcock.

(2) Attach a rubber adapter to a Buchner funnel and set the funnel at the top of the filtration chamber.

(3) Put a receiving container in the filtration chamber. If the container is too far from the tip of the funnel, place a rubber board underneath.

(4) Wet the contact between the bottom rim of the filtration chamber and the coaster glass plate. Slightly rubbing their ground glass faces to obtain a good contact.

(5) Put a filter paper slightly smaller than the inside of the funnel and add a small amount of a solvent to wet the filter paper.

(6) Turn on the aspirator and open the three-way stopcock to fit the filter paper to the funnel.

Chapter 3   Data and Error Analysis

Performing the experiment and collecting data are only the beginning of the process of completing an experiment in science. Understanding the results of any given experiment is always the central goal of the experiment. Presenting those results in a clear concise manner completes the experiment. This overview of the complete process is as valid in an instructional laboratory course as in a research environment.

1. Error

The words "error" and "uncertainty" are used to describe the same concept in measurement. It is unfortunate that the term, "error" is the standard scientific word because usually there is no mistake or error in making a measurement. Frequently, the uncertainties are dominated by natural irregularities or differences in what is being measured.

Errors may arise from three sources:

(1) Careless errors: These are due to mistakes in reading scales or careless setting of markers, etc. They can be eliminated by repetition of readings by one or two observers.

(2) Systematic errors: These are due to built-in errors in the instrument either in design or calibration. Repetition of observation with the same instrument will not show a spread in the measurements. They are the hardest source of errors to detect.

(3) Random errors: These always lead to a spread or distribution of results on repetition of the particular measurement. They may arise from fluctuations in either the physical parameters due to the statistical nature of the particular

phenomenon or the judgment of the experimenter, such as variation in response time or estimation in scale reading.

For each measured value A, there is an estimated error ΔA. The complete result is given by $A \pm \Delta A$. This means that the "true value" probably lies between a maximum value of $A+\Delta A$ and a minimum value of $A-\Delta A$. Sometimes the terms relative error and percent error are used, where:

$$\text{Relative Error} = \frac{\text{Estimated Error}}{\text{Estimated Value}} \times 100\% = \frac{\Delta A}{A}\%$$

2. Accuracy and Precision

Firstly, When one considers the quality of a measurement there are two aspects to consider. The first is if one were to repeat the measurement, how close would new results be to the old, i.e., how reproducible is the measurement? Scientists refer to this as the precision of the measurement.

Secondly, a measurement is considered "good" if it agrees with the true value. This is known as the accuracy of the measurement. But there is a potential problem in that one needs to know the "true value" to determine the accuracy.

3. Significant Figures

Measured quantities are often used in calculations. The precision of the calculation is limited by the precision of the measurements on which it is based. The concept of significant figures is introduced to define the real precision of measurements.

The rules for identifying significant digits when writing or interpreting numbers are as follows:

(1) Non-zero digits are always significant.

(2) All zeros between other significant digits are significant.

The number of significant figures is determined starting with the leftmost non-zero digit. The leftmost non-zero digit is sometimes called the most significant digit or the most significant figure.

For example, in the number 0.004205, the "4" is the most significant digit. The left-hand '0's are not significant. The zero between the "2" and the "5" is significant.

The rightmost digit of a decimal number is the least significant digit or least significant figure. Another way to look at the least significant figure is to consider it to be the rightmost digit when the number is written in scientific notation. Least significant figure is still significant! In the number 0.004205 (which may be written as 4.205×10^{-3}), the "5" is the least significant figure. In the number 43.120 (which may be written as 4.3210×10), the "0" is the least significant figure.

If no decimal point is present, the rightmost non-zero digit is the least significant figure. In the number 5800, the least significant figure is "8".

Rules When Significant Digit Involves in Calculation

1. Addition and Subtraction

When measured quantities are used in addition or subtraction, the uncertainty is determined by the absolute uncertainty in the least precise measurement (not by the number of significant figures). Sometimes this is considered to be the number of digits after the decimal point.

Example: A: 32.01 m; B: 5.325 m; C: 12 m.

When added the previous data together (A+B+C), you will get 49.335 m, but the sum should be reported as "49" m.

2. Multiplication and Division

When experimental quantities are multiplied or divided, the number of significant figures in the result is the same as that in the quantity with the smallest number of significant figures. For example, a density calculation is made in which 25.624 grams is divided by 25 mL, the density should be reported as 1.0 g/ mL, not as 1.0000 g/ mL or 1.000 g/ mL.

Part 2 Experiments

Experiment 1 Determination of the Ideal Gas Constant

I. Objects

(1) Ensure to grasp the application of ideal gas law and law of partial pressure.

(2) Understand the factors that affect accuracy of determination results of ideal gas constant.

(3) Learn the method of determination of ideal gas constant and practice the procedures.

II. Principles

The Ideal Gas Law could represent by following formula:

$$pV = nRT$$

Here, p represents as the gas pressure (in atmospheres);

  V——The gas volume (in Liter);

  n——The number of moles of gas in the sample;

  T——The gas temperature (in Kelvin).

  R——A proportionality constant called the Gas Constant, and has a theoretical value of 0.08206 L·atm/(K·mol).

The Ideal Gas Law is formulated only for ideal gases, some postulates are introduced. Such as gases are very tiny particles that are in constant and rapid motion. They have completely elastic collisions in which no energy is lost and there is no interaction between the particles. In practice, no gas completely obeys the Kinetic Molecular Theory. Most gases have some size. The larger the gas molecule, the less closely it follows the Ideal Gas Law. In general, some small gases such as H_2, O_2 and N_2 obey the ideal gas law approximately under atmosphere pressure and room temperature. It could be calculated the gas constant by determining p, V and n value experimentally.

This experiment uses the single displacement reaction between aluminum metal and hydrochloric acid which could generate hydrogen gas under atmosphere pressure and room temperature. To determine gas constant by measuring H_2 is quite reasonable since it shows less error compare to the ideal gas. The reaction formula is

$$2Al + 6HCl \Longleftrightarrow 2AlCl_3 + 3H_2\uparrow$$

Students will then obtain the following values for the collected sample of hydrogen gas: ① Volume; ② Temperature; ③ Moles; ④ Pressure. Measuring a certain amount of Al to react with overdose HCl under certain pressure and temperature. The hydrogen volume will be directly measured from the eudiometer scale, and the apparatus is shown in Fig. 2-1a. The hydrogen temperature will also be directly measured using a thermometer. The numbers of mole of H_2 ($n(H_2)$) of the collected hydrogen can be easily calculated from the measured mass of the Al reactant using stoichiometry.

$$R = p(H_2)\, V(H_2) / n(H_2)\, T$$

In which

$$n(H_2) = 3\, W_{Al} / 2M_{Al}$$

W_{Al}: Mass of Al.
M_{Al}: Relative atomic mass of Al.

$$V(H_2) = V_2 - V_1$$

V_1: Volume of eudiometer before reaction.

V_2: Volume of eudiometer after reaction.

However, the mole quantity and pressure of the hydrogen gas must be determined indirectly. But the hydrogen pressure is a little more difficult to obtain. Since hydrogen is collected over a water bath, a small amount of water vapor is mixed with the hydrogen in the eudiometer. The combined pressure of the H_2 and H_2O gases will be equal (after adjustments) to the external atmospheric pressure.

$$p(H_2) = p_{atm} - p_{water}$$

p_{atm} (atmospheric pressure) will be measured using a barometer.

p_{water} (the partial pressure of water vapor) depends on the temperature of the water bath, and can be obtained from the Table supplied in appendix 6.

III. Apparatus and Reagents

1. Apparatus

Apparatus Name	Item Mode	Specification
Set up for determining gas constant	shown in Fig.2-1b	
Electronic balance		1 kg
Thermo barometer	1	
Graduated cylinder	1	10 mL
Beaker	3	100 mL
Dropper	1	20 mm
Sandpaper (fine)		
Slice of filter paper		20 mm × 100 mm

2. Reagents

Reagents Name	Status	Mass or concentration
Copper wire	Solid	
Hydrochloric acid	Solution	6 mol·L^{-1}
Aluminum ribbon	Solid	0.023~0.030 g (measured by students)

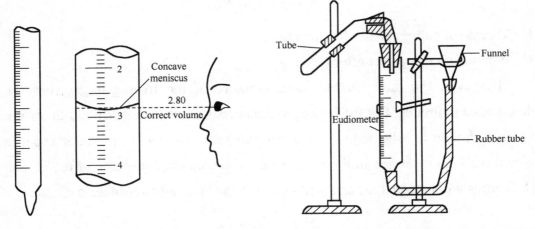

Fig. 2-1a The rules of reading the concave meniscus Fig. 2-1b Set up for determining gas constant

IV. Procedures

(1) Make sure the apparatus of determining gas constant is settled as described in Fig. 2-1b.

(2) Add some water to the eudiometer till the water level under "0.00" mark. Move the rubber tube up and down rapidly in order to get rid of bubbles.

(3) Before reaction, the gas tightness has to be checked first. Block the tube with a stopper then move the tube up and down for many times, and observe the status of water level. If the water level is not stable, a gas leakage exists somewhere in the set up. Check all the joint spots, and repeat the previous procedure until the water level is stable.

(4) Measure 5 mL of 6 mol·L^{-1} HCl by graduated cylinder, and drop the HCl solution into reaction tube carefully. Make sure that no HCl drops remain on the upper part of tube (Why?). If there are some HCl remains, wipe them by slice of filter paper.

(5) Weight the Al ribbon and record it's mass. Wrap the Al around the end of the copper wire and place the wrapped Al ribbon on the horizontal part of tube (Make sure that the wrapped Al ribbon will not fall into the tube, and considerate the reason for it?).

(6) Check the gas tightness once again. If there is no gas leakage, the pressure equalization has to be checked. It means the level of the water inside the tube must be the same as the level of water outside the tube. To achieve this, internal and external water levels are equal. Then read the V1 after 5 minutes (The rules of reading the concave meniscus is shown in Fig. 2-1a).

(7) Lightly vibrate the reaction tube in order to slip the wrapped Al ribbon into HCl solution. During the reaction, gas pressure will increase since H_2 is generated. Thus it is better to reduce gas pressure to prevent gas leakage. In practical, the easiest and effective move is keeping the water level in the eudiometer equal to the water level in the funnel by adjusting the rubber tube down.

(8) Since this reaction is exothermic, the fresh generated H_2 is not under room temperature. The final volume of H_2 is an essential factor when considerate less experiment error. While water level of eudiometer is not decreasing, move the rubber tube to make the two water levels equal to each other and record the concave meniscus. Then repeat previous adjusting procedure again until the difference of two observed concave meniscus values is less than 0.05 mL. We could determine gas temperature is almost equal to room temperature. The final H_2 volume is obtained as V_2.

(9) Determine the room temperature and p_{atm} by using thermo barometer. p_{water} (the partial pressure of water vapor) depends on the temperature of the water bath, and can be obtained from the appendix 6.

V. Data Recording and Analyzing

1. Data recording

Mass of Al: $W_{Al} = $ _____ g.

Volume of eudiometer before reaction: $V_1 = $ _____ mL = _____ m^3.

Volume of eudiometer after reaction: $V_2 = $ _____ mL = _____ m^3.

Room temperature: $t = $ _____ °C = _____ K.

Atmosphere pressure: $p_{atm} = $ _____ kPa = _____ Pa.

The partial pressure of water vapor at room temperature: $p_{water} = $ _____ Pa.

2. Data Analyzing

Volume of H_2: $\quad V(H_2) = V_2 - V_1 = $ _____ m^3.

Partial pressure of H_2: $\quad p(H_2) = p_{atm} - p(H_2O) = $ _____ Pa.

Numbers of moles of H_2: $\quad n(H_2) = 3W_{Al}/2M_{Al} = $ _____ mol.

Ideal gas constant: $\quad R_{real} = p(H_2)V(H_2)/n(H_2)T = $ _____ $J \cdot mol^{-1} \cdot K^{-1}$.

Ideal gas constant in theory: $\quad R_{theory} = 8.314 \; J \cdot mol^{-1} \cdot K^{-1} (Pa \cdot m^3 \cdot mol^{-1} \cdot K^{-1})$

$$= 0.082 \; atm \cdot L \cdot mol^{-1} \cdot K^{-1}$$

$$= 62400 \; mmHg \cdot mL \cdot mol^{-1} \cdot K^{-1}.$$

Relative Error: $\quad |R_{theory} - R_{real}|/R_{theory} \times 100\% = $ _____.

Notice: The relative error of this experiment is required less than 5%. If it is higher than 5%, you should do the experiment again at once.

VI. Discussion and Exercises

(1) Is the volume of H_2 determined by experiment equal to the value of dry H_2 gas volume under the same temperature and pressure?

(2) When calculate mole number of H_2, why not considerate air partial pressure since some air was existed in the eudiometer?

(3) Discuss how it affect the ideal gas constant under the following conditions:

a) Bubbles exist in the eudiometer.

b) Read the volume of V as the gas is still hot.

c) Gas leakage exists in the set up.

d) Oxidation film cover the Al ribbon.

e) Water overflow out through the funnel since the pressure in eudiometer is too high.

f) The mistaken of mass measurement of Al ribbon.

Attachment

1. Operation of Thermo Barometer

Thermo barometer is an instrument which is applied for measuring temperature and pressure at the same time. It contains two sensors related to temperature and pressure respectively. The range of temperature is from 0℃ to 99.99℃, resolution is 0.01℃. The range for pressure is 101.3 kPa±20 kPa, resolution is 0.1 kpa.

Operation procedures:

(1) Place thermo barometer on a platform with less air disturbance.

(2) Switch on the thermo barometer, read the temperature value and pressure after it is stabilized.

(3) Pay attention that non water or substance could enter through input port. Do not fold it.

(4) Please do not put any substance on thermo barometer preventing high load pressure and chemical corrosion.

2. Electronic Balance

Electronic balance is a precise weighing instrument which measuring mass of substance of any form. It should be handled carefully. Especially, a sample should be placed softly on the balance pan.

Some points have to be checked before using the balance:

The range of this balance is 0 g to 200 g. The error is 0.0002 g. Before use, the balance is confirmed to be placed horizontally with the spirit level. When it is not horizontal, it should be adjusted by turning the level screw.

(1) Close the sliding glass doors, and turn the balance on (Press <ON>). You will see :

| ± 8888888 % |
| O   g |

(2) The self-check process is scanning. The balance mode is shown after it is stabilized for 2 seconds:

| --2004-- |

(3) Then the weighing mode:

| 0.0000 g |

(4) Press <OFF>, the screen is switch off. If it is not going to use for a long

time, make sure to plug out the power sources.

(5) You should take hold of the container with a folded weight paper in two and be careful not to touch it with your bare hands. Open one of the sliding glass doors, put the container softly on the center of the balance pan, and close the door. The weight of the container will shown:

> 18.9001 g

(6) Press the tare button and display will again read 0.0000 and wait for the mark on the left to come out. This is the stability indicator, indicating that the weight is stable. This operation is called as canceling tare.

> 0.0000 g

When you take out the container, it will show a negative value:

> −18.9001 g

Press<TAR>again, it will turn out to 0.00 again:

> 0.0000 g

Electronic Balance Calibration

In order to make sure the measurement accuracy, a calibration has to be taken after long time period or any modification of balance position.

First, take out all the objects from the balance pan, put COU—0, UNT—g, INT—3, ASD—2 mode, press <TAR>. It returns to 0.

Press <CAL>, do not touch the button after it shows "CAL—". Then "CAL—200" display in the screen which means calibration using 200 g standard balance weight. Now put the standard balance weight of 200 g in the center of balance pan, a waiting mode is processing shown as "-------" .Wait several seconds until it shown "200.000 g", then it should show "0.000 g" after taking away the standard balance weight of 200.00 g. Repeat the previous process if the screen does not show "0.000 g". The calibration process is under this order:

Experiment 2 Determination of Enthalpy of Chemical Reaction—Zn and CuSO₄ Reaction

I. Objects

(1) To determine the enthalpy change for the single displacement reaction between zinc and copper sulphate.

(2) Grasp the method of extrapolation to determine the temperature change during the reaction.

II. Principles

Heat is associated with nearly all chemical reactions. In such instances, the reaction either liberates heat (exothermic) or absorbs heat (endothermic). When a reaction is carried out under constant pressure (as in an open beaker), the heat associated with the reaction is known as enthalpy. The symbol for enthalpy change is ΔH. It is most often too difficult to directly measure the enthalpy change for a reaction. What can be done is to measure the heat change that occurs in the surroundings by monitoring temperature changes. Conducting a reaction between two substances in aqueous solution, allows the enthalpy of the reaction to be indirectly calculated with the following equation.

The reaction between Zn and CuSO₄ is a spontaneous exothermic reaction, the enthalpy under 298.15 K (ideal) is 216.8 kJ per mole.

$$Zn + CuSO_4 = ZnSO_4 + Cu$$

$$\Delta_r H_m^\ominus = -216.8 \ (kJ \cdot mol^{-1})$$

This experiment is to add an excess of zinc powder to a known amount of copper(II)sulphate solution, and measuring the temperature change over a period of time, the enthalpy change for the reaction could be calculated by the Hess's law.

$$Q = \Delta T C d V + \Delta T C_p \tag{1}$$

On the assumption that the heat absorbed by the container $\Delta T C_p$ could be ignored:

$$\Delta_r H_m^\ominus = -Q/n = -\Delta T C d V / 1000 n \tag{2}$$

In which, Q——The heat change during the reaction between Zn and $CuSO_4$(kJ);

$\Delta_r H_m^\ominus$——Enthalpy of the reaction (kJ·mol^{-1});

ΔT——Temperature change during the reaction (K);

C——Specific heat of the solution, which could be substituted by specific heat of water at 298 K (4.18 kJ·mol^{-1});

d——Density of solution, which could be substituted by water density (1.00 g·mL^{-1});

n——Mole number of $CuSO_4$ solution;

1000——Convertion factor.

On the basis on the formula (2), the enthalpy could be calculated by using molar concentration of $CuSO_4$ and the temperature change.

III. Apparatus and Materials

1. Apparatus

Apparatus Name	Item Mode	Specification
Chemical heat of formation tester	CXJ-2	
Balance		1 kg
Transfer Pipette		50 mL
Aurilave		

2. Materials

Materials Name	Status	Amount
$CuSO_4$	solution	0.2 mol·L^{-1}
Zn	powder	3 g

The Chemical reaction heat tester (Mode CXJ-2) is shown in Fig. 2-2.

Fig. 2-2 Chemical reaction heat tester (Mode CXJ-2)

IV. Procedures

(1) Open the tester, and take out of the insulated cup. Washing the cup carefully and wipe it by filter paper.

(2) Pipette 100 mL of $CuSO_4$ solution into the insulated cup.

(3) Put the cup into the tester and close the cover. Switch on the tester begin to stir the solution.

(4) Press the time recorder button, and record the temperature every half minute. Record the beginning temperature T_1 until the temperature keeps being stable for at least 2 minutes.

(5) Put the Zn powder into the reactor through the small hole. Record the temperature every quarter minute. As the temperature arrives the maximum value (reaction ending temperature) T_2, keep on recording the temperature each half minute until gathering at least 6 points.

(6) Take out the insulated cup, and pour out the wasted mixture to the waste

liquid bottle. Wash the cup and keep it in order. Leave the laboratory after the instructor checks the experimental data.

V. Data Recording and Analyzing

1. Data Recording

CuSO$_4$ concentration_____mol·L^{-1}, amount_____mL, Zn powder_____g.

The relation between reaction time and temperature is shown in the Tab. 2-1.

Tab. 2-1 The relation between reaction time and temperature

Time/min						
Temperature/℃						
Time/min						
Temperature/℃						

2. Data Analyzing——Determine ΔT by Extrapolation

Since it could not achieve absolute heat insulation during the experiment, heat exchange should occur between reactor and surroundings. The experimental T_2 is not accurate reaction ending temperature. Here use extrapolates curve method to determine ΔT by the following procedures.

(1) Plot the temperature (vertical axis) against time (horizontal axis).

(2) Extrapolate the post reaction plots curve back to left to establish the maximum temperature rise shown in Fig. 2-3.

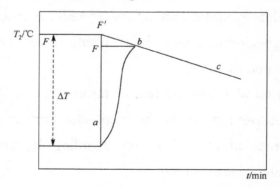

Fig. 2-3 Determination of ΔT by extraplation method

(3) a represent the reaction beginning point, and b is the observing reaction ending point.

(4) Real temperature change during reaction could represent by: $\Delta T = T_2 - T_1$.

(5) Then calculate $\Delta_r H_m^{\ominus}$ by using ΔT.

3. Error Analysis

Using the following formula to analyze experiment relative error:

$$\text{Relative error} = [(\Delta_r H_{m\ \text{theory}}^{\ominus} - \Delta_r H_m^{\ominus}) / \Delta_r H_m^{\ominus}] \times 100\%$$

If the relative error is greater than 10%, please analyze the reason.

VI. Discussion and Exercises

(1) Why not use electron balance for measuring Zn powder?

(2) Why not use graduated cylinder for measuring $CuSO_4$ solution?

Experiment 3 Chemical Reaction Rate

I. Objects

(1) Know the approach of determining chemical reaction rate.

(2) Understand the factors that affect reaction rate, such as concentration, temperature and catalyst.

(3) Calculate the activate energy of reaction by plotting slope.

II. Principles

In an acidic solution, there is a reaction between $KBrO_3$ and KI:

$$6KI + KBrO_3 + 6HCl = 3I_2 + KBr + KCl + 3H_2O$$

The net ionic formula could represent as following:

$$6I^- + BrO_3^- + 6H^+ = 3I_2 + Br^- + 3H_2O \quad (1)$$

The reaction rate could represent as following:

$$v = \frac{\Delta c(BrO_3^-)}{\Delta t} = kc^x(BrO_3^-)c^y(I^-)$$

In which, v ——The average rate of reaction[①];

$\Delta c(BrO_3^-)$ ——The concentration change of BrO_3^- during a time period Δt;

$c(BrO_3^-)$ and $c(I^-)$ ——The initial concentration of BrO_3^- and I^- ($mol \cdot L^{-1}$);

[①] The reactant concentration variation is quite small ($\approx 10^{-4} mol \cdot L^{-1}$) during reaction time Δt. It may not introduce too big error since using average rate of reaction instead of instantaneous rate.

k —— Rate constant;

$x+y$ —— Value of the sumption of two exponents reaction order[①].

Based on previous discussion, it is reasonable to calculate of rate constant k and reaction order $(x+y)$ by determining variation of BrO_3^- concentration during Δt.

1. Determining Rate Constant

Since I_2 is generated as the reaction occurs, it is too hard to determine BrO_3^- concentration variation by using starch as the indicator of in a certain time period Δt. In order to measure BrO_3^- concentration variation value, a mixture of sodium thiosulfate ($Na_2S_2O_3$ solution) and starch is added to KI solution first. Then mix that with HCl-treated $KBrO_3$ solution. Then $Na_2S_2O_3$ solution could react with the I_2 as shown by the following reaction equation.

$$2S_2O_3^{2-} + I_2 \rightleftharpoons S_4O_6^{2-} + 2I^- \tag{2}$$

The I_2 could react with $S_2O_3^{2-}$ to form colorless $S_4O_6^{2-}$ and I^- immediately since reaction rate of reaction (2) solution is much higher that reaction (1). If this will happen, the solution will not show blue color which are resulted from the reaction I_2 and starch at the first period of reaction. Once $Na_2S_2O_3$ is totally consumed, the solution is going to become blue since tiny amount of I_2 is detected by starch. Stoichiometries tell us, once 1 mol BrO_3^- is consumed in reaction (1), 1 mol $S_2O_3^{2-}$ has to be consumed in reaction (2). The relationship between the concentration could represent as following:

$$\Delta c(BrO_3^-) = \frac{\Delta c(S_2O_3^{2-})}{6}$$

$\Delta c(S_2O_3^{2-})$ could represent initial concentration of $Na_2S_2O_3$ solution since the concentration of $S_2O_3^{2-}$ is turning to 0 which means it is totally consumed during

---

[①] If takes hydrogen ion concentration into account in the reaction, the reaction rate express as following:

$$v = \frac{\Delta c(BrO_3^-)}{\Delta t} = kc^x(BrO_3^-)c^y(I^-)c^z(H^+)$$

In this experiment we assume that reaction rate does not depend on hydrogen ion concentration as regard as H^+ is settled to a certain value.

Δt. Therefore, we could use initial concentration of Na$_2$S$_2$O$_3$ solution as a substitute of $\Delta c(S_2O_3^{2-})$. The relationship between those concentrations could represent as:

$$\frac{\Delta c(BrO_3^-)}{\Delta t} = \frac{\Delta c(S_2O_3^{2-})}{6\Delta t} = kc^x(BrO_3^-)c^y(I^-)$$

Then

$$k = \frac{\Delta c(S_2O_3^{2-})}{6\Delta t c^x(BrO_3^-)c^y(I^-)} \tag{3}$$

2. Reaction Orders

The reaction rate could be determined since different $c(I^-)$ and $c(BrO_3^-)$ is introduced and fixed at a certain temperature. The value of two exponents (x, y) could be calculated by using the following formula and the sum of x and y is called reaction order.

$$\frac{v_1}{v_2} = \frac{kc^x(BrO_3^-)1c^y(I^-)}{kc^x(BrO_3^-)2c^y(I^-)} = \frac{c^x(BrO_3^-)_1}{c^x(BrO_3^-)_2}$$

Since $\dfrac{v_1}{v_2} = \dfrac{\Delta t_2}{\Delta t_1}$, it could obtain:

$$x = \frac{\ln\dfrac{\Delta t_2}{\Delta t_1}}{\ln\dfrac{c(BrO_3^-)_1}{c(BrO_3^-)_2}} \tag{4}$$

The value of y could be calculated by the similar procedure described above by varying $c(I^-)$ and fixing $c(BrO_3^-)$. The sum of x and y is the reaction order.

In this experiment, the x and y value is already determined as $x = 1$, $y = 1$.

3. Activation Energy

The reaction rate k is related to the temperature of system by what is known as arrhenius equation:

$$\lg k = \frac{-E_a}{2.303RT} + A \qquad (5)$$

In which, E_a——Activation energy of reaction ($J \cdot mol^{-1}$);

R——Ideal gas constant ($R = 8.3145 \, J \cdot mol^{-1} \cdot K^{-1}$);

T——Temperature in degrees Kelvin (K);

A——Constant, frequency factor (A is a stable value for a certain reaction).

In this experiment just measure the rate constant at several temperature points, plot lg k and $1/T$, then activation energy is determined from the slope. Since the intercept of the slope equals to $-E_a/2.303R$, which could be determined as value of a/b (Fig. 2-4). Then it means:

$$\frac{a}{b} = \frac{-E}{2.303R} \quad \text{or} \quad E = -\frac{a}{b} \times 2.303R \qquad (6)$$

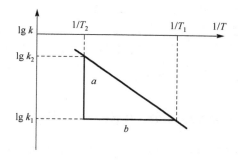

Fig. 2-4 Determine activation energy by ploting lg k and $1/T$

III. Apparatus and Reagents

1. Apparatus

Apparatus Name	Specification	Quantity
Test tubes	25 mL	6
Graduate cylinder	10 mL	6
Beaker1	150 mL	1
Beaker2	50 mL	1
Stirrer rod	150~200 mm	1

Continued

Apparatus Name	Specification	Quantity
Electric magnetic stirrer		1
Stopwatch		1
Digital constant temperature water bath		3
Glass thermometer	0~100℃	1

2. Reagents

Reagents Name	Concentration/mol·L^{-1}
KI solution	0.01
KBrO$_3$ solution	0.04
Na$_2$S$_2$O$_3$ solution	0.001
HCl solution	0.1
KNO$_3$ solution	0.01
KNO$_3$ solution	0.04
(NH$_4$)Mo$_7$O$_{24}$·H$_2$O solution	0.06
Starch-iodide indicator	0.2%

IV. Procedures

1. The Effect of Concentration on Reaction Rate

Prepare two solutions at room temperature. The first one is a mixture of 10 mL KI solution at concentration of 0.01 mol·L^{-1}, 10 mL Na$_2$S$_2$O$_3$ solution at concentration of 0.001 mol·L^{-1} and 2 mL starch indicator at concentration of 0.2%. Place this mixture into a 150 mL beaker and stirring vigorously with electric magnetic stirrer. Then the second solution is measuring 10 mL KBrO$_3$ solution at concentration of 0.04 mol·L^{-1} and 10 mL HCl solution at concentration of 0.1 mol·L^{-1} by graduated cylinder which is placed into 50 mL beaker. Shake the beaker gently and quickly pour it into the first solution with stirring. Record starting time by stopwatch when the two solutions mixed. Record the ending time once blue color appearances. Switch off the electric magnetic stirrer and write down the time in the first column. Repeat the same previous procedure according to different concentration values of each reagent in Tab. 2-2 (column 2 and 3). In this experiments we use a KNO_3 solution to make sure the K^+ concentration remain constant.

2. The Effect of Catalyst on Reaction Rate

Repeat the same procedure to prepare solution 1 and 2 as described previous according to Tab.2-2 (column 1). Just remember to add an extra drop of $(NH_4)Mo_7O_{24} \cdot H_2O$ solution to solution 1. Then mix the two solutions to observe the color changing point and compare the results with the one solution without catalyst.

Tab. 2-2 Reaction rate influenced by amount of concentration and catalyst

	Experiment Number	1	2	3	4
Amount/mL	KI solution (0.01 mol·L^{-1})	10	5	10	5
	KBrO$_3$ solution (0.04 mol·L^{-1})	10	10	5	10
	Na$_2$S$_2$O$_3$ solution (0.001 mol·L^{-1})	10	10	10	10
	HCl solution (0.1 mol·L^{-1})	10	10	10	10
	Starch solution (0.2%)	2	2	2	2
	KNO$_3$ solution (0.01 mol·L^{-1})	0	5	0	5
	KNO$_3$ solution (0.04 mol·L^{-1})	0	0	5	0
	(NH$_4$)$_6$Mo$_7$O$_{24}$·4H$_2$O solution (0.06 mol·L^{-1})	0	0	0	1 drop
Initial concentration /mol·L^{-1}	KI solution				
	KBrO$_3$ solution				
	Na$_2$S$_2$O$_3$ solution				
Reaction Time t/s					
Reaction Rate k					

3. The Effect of Temperature on Reaction Rate

(1) Prepare two 25 mL beakers, the first one named 1st test beaker is filled with 5 mL KI solution (0.01 mol·L^{-1}) and 2 mL starch solution (0.2%); the second one (2nd test beaker) contains a mixture of 5 mL KBrO$_3$ solution with a concentration of 0.04 mol·L^{-1} and 5 mL HCl solution (0.1 mol·L^{-1}).

(2) Pour the solution in 2nd test tube into 1st test beaker and record starting time

immediately after mixing at room temperature. Stirring mixture continually and record the ending time as the first blue color is observed in the solution.

(3) Repeat the same procedure at three more temperature points in the constant temperature water bath which is above room temperature 10℃, 20℃ and 30℃ respectively. An attention has to be taken that the both two test beakers have to remain in the constant temperature water bath for 5 to 8 minutes to achieve the desired temperature. Then pour the solution of 2nd test beaker into 1st test beaker and stirring continually to watch color change. Record time and temperature.

(4) Put the four experimental data into Tab. 2-3 and calculate k, $\lg k$ and $1/T$.

Tab. 2-3 The effect of temperature on the chemical reaction rate

	Experiment Number	1	2	3	4
	Temperature	Room temperature(r.t.)	r.t.+10℃	r.t.+20℃	r.t.+30℃
Amount/ mL	KI solution (0.01 mol·L^{-1})	5	5	5	5
	KBrO$_3$ solution (0.04 mol·L^{-1})	5	5	5	5
	Na$_2$S$_2$O$_3$ solution (0.001 mol·L^{-1})	5	5	5	5
	NaHSO$_4$ solution (0.1 mol·L^{-1})	5	5	5	5
	Starch solution (0.2%)	2	2	2	2
Initial concentration /mol·L^{-1}	KI solution				
	KBrO$_3$ solution				
	Na$_2$S$_2$O$_3$ solution				
Reaction Time t/s					
Reaction Rate k					

4. Determine Activation Energy by Plotting a Slope

(1) Plot a slope by using $1/T$ as x axis and $\lg k$ value as y axis.

(2) Determine the intercept value (a/b) from the slope.

(3) From formula (6), intercept $= a/b = -E/2.303R$, $E = -2.303 \times R \times (a/b)$ kJ·mol^{-1}.

V. Discussion and Exercises

(1) Could the reaction formula express as $v = kc(BrO_3^-)c^6(I^-)$ based on law of mass action?

(2) What are the different influences to reaction rate by concentration and temperature?

(3) Does the concentration of reagents during reaction equal to the initial concentration?

(4) Why could determine the ending time of reaction when the solution color changes? Does it mean that the reactions terminate as the first blue color shows?

(5) Does following condition affect reaction rate? If it affects, what consequences will bring?

a) Measure different 6 reagents using the same graduated cylinder;

b) Do not add HCl solution into the solution or add less;

c) Add the $KBrO_3$ solution mixture into the first solution slowly;

d) With out stirring during reaction.

Experiment 4 Electrochemistry

I. Objects

(1) Understand the composition of the galvanic cell and the rough determination of its potential.

(2) Understand the application of electrolysis principle —— electroplating.

(3) Understand the basic principles and methods of the electrochemical corrosion and protection of metal.

II. Principles

1. Galvanic Cell

A setup through which chemical energy of a redox reaction can convert into electrical energy is called galvanic cell. Generally, a galvanic cell consists of two electrodes, electrolyte solution and a salt bridge. In a galvanic cell, the oxidation reaction and the reduction reaction were carried out in anode and cathode, respectively. Electrons flow out from anode and flow into the cathode through the out circuit. If we connect a voltmeter between two electrodes, then the voltage of this galvanic cell can be measured roughly. Namely, the electric potential of the galvanic E, $E = \varphi_+ - \varphi_-$.

2. Electroplating——Electroplating Copper on Iron Nail

Electroplating is a process that depositing metal on the surface of another metal electrode which are cathode in a setup where contains electrolyte solutions driving by the out direct current power. In order to improve the corrosion resistance, in industry, the technology of plating chromium on steel is more often used. The main purpose of plating copper on iron is to combine the substrate and the surface coating of metal very well using the copper as interlayer.

In order to get a plating layer with good quality, firstly, the surface of the metal should be oil removal, dust removal. In addition, temperature and current density should also be well controlled.

The compositions of the electrolysis solution we choose include: oxalic acid ($H_2C_2O_4$ solution), ammonia and copper sulphate ($CuSO_4$ solution). The $H_2C_2O_4$ solution and $NH_3 \cdot H_2O$ solution can react with $CuSO_4$ solution to form a coordination compound $(NH_4)_4[Cu(C_2O_4)_3]$ solution (ammonium copper oxalate), then, the iron Cu^{2+} with a proper concentration can be obtained by the dissociation of the coordination ion $[Cu(C_2O_4)_3]^{4+}$. The detailed reactions are as following:

$$CuSO_4 + 4NH_3 \cdot H_2O \rightleftharpoons [Cu(NH_3)_4]SO_4 + 4H_2O$$

$$[Cu(NH_3)_4]SO_4 + 3H_2C_2O_4 \rightleftharpoons (NH_4)_4[Cu(C_2O_4)_3] + H_2SO_4$$

$$[Cu(C_2O_4)_3]^{4-} \rightleftharpoons Cu^{2+} + 3C_2O_4^{2-}$$

During the electroplating process, Cu^{2+} obtained electron in cathode and be reduced to Cu atom following by depositing on cathode.

In the electrolysis solutions with cooperation ions, the concentration of free metal ions is low so that the plated coatings are fine and uniform, which combined closely with the substrate and is not easy to peel off.

3. Electrochemical Corrosion And Protection of Metal

Generally, due to the inhomogeneity of composition or any other affection, there are different areas with different electrode potentials. When electrolyte is existed on this metal surface, a corrosion cell can be formed and make the metal surface attainted. This phenomenon is called electrochemical corrosion of metal. In the corrosion cell, the more active metal always is the anode which is oxidized and eroded. However, the cathode just transfers the electrons, resulting in the reduction of H^+ or O_2, so the cathode was not eroded.

A redox reaction may happen between Zn and HCl solution and H_2 was emitted. However, the corrosion rate was different when a galvanic cell was formed or not. The difference of emitting rate of H_2 can be observed by comparing the experiments of Zn and HCl solution or Cu-Zn and HCl solution. When the Zn plating coating in Zn coated Fe was destroyed, which metal will be eroded? The answer can be obtained by an experiment using $K_3[Fe(CN)_6]$ solution as indicator. If the Fe was eroded, then the resulting Fe^{2+} will react with $[Fe(CN)_6]^{3-}$ and form a blue sedimentation.

$$3Fe^{2+}+2[Fe(CN)_6]^{3-}=\!=\!=Fe_3[Fe(CN)_6]_2 \downarrow \text{ (blue sedimentation)}$$

If Zn was eroded, then the resulting Zn^{2+} will react with $[Fe(CN)_6]^{3-}$ and form a light yellow sedimentation.

$$3Zn^{2+}+2[Fe(CN)_6]^{3-}=\!=\!=Zn_3[Fe(CN)_6]_2 \downarrow \text{ (light yellow sedimentation)}$$

The reagent being added into the corrosion system which can terminate or delay the corrosion process was called corrosion inhibitor. For example, urotropine or aniline can be used as corrosion inhibitors in metal corrosion under acid medium.

If an external DC (Direct Current) power was supplied and connected the metal which will be protected into cathode of the power, the power will supply electrons, and the potential of this metal maybe decreased. Therefore, the corrosion to this metal can be avoided, the method was called as cathode protection method.

III. Apparatus and Reagents

1. Apparatus

Apparatus Name	Specification	Number
Voltmeter (Public)	0~3 V	Several
Thermometer (Public)	0~100 ℃	1
File (Public)		Several
Glass tubes	10 mL	6
Shelf for tube		1

Continued

Apparatus Name	Specification	Number
Drop board		1
Electroplating bottle		1
Salt bridge		1
Sandpaper		Several
Filter paper strips		Several
Power cord (with wire hook and metal clip)		2
Copper electrode board (with wiring terminal)		1
Zinc electrode board (with wiring terminal)		1
Zinc-coated iron		1
Small Nails		3
Large Nails		1
Copper rod	$\Phi 1 \times 200$ mm	1
Zinc particle	$\Phi 2 \sim 3$ mm	Several

2. Reagents

Reagents Name	Concentration	Reagent Name	Concentration
$ZnSO_4$ solution	1 mol·kg^{-1}	NaCl solution	1 mol·kg^{-1}
$CuSO_4$ solution	1 mol·kg^{-1}	$K_3[Fe(CN)_6]$ solution	0.1%
HCl solution	0.1 mol·kg^{-1}	Phenolphthalein	
HCl solution	1 mol·kg^{-1}	Urotropine	
		Electrolyte	

IV. Procedures

1. Assemble a Galvanic Cell and Measure its Potential Roughly

A Cu-Zn galvanic cell can be assembled according to the schematic diagram on Fig. 2-5. Firstly, zinc piece which has been polished using sandpaper was inserted the into $ZnSO_4$ solution, similarly, copper piece was inserted into $CuSO_4$ solution. Then, two solutions were connected using KCl salt bridge. Zinc piece and copper piece were connected with the cathode and anode of the Voltmeter, respectively. Here, a Cu-Zn galvanic cell was obtained, you should measure its voltage roughly, write the corresponding electrode reaction, comparing the difference between experimental data and theoretical value and explain why. What's the role of the salt bridge?

After the experiment was completed, salt bridge should be rinsed and return to the KCl solution.

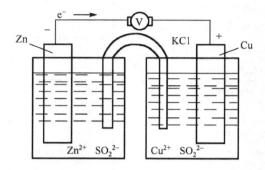

Fig. 2-5　Schematic diagram of galvanic cell

2. Electroplating——Electroplating Copper on Iron Nail

Pretreatment of the iron nail: a big iron nail was polished using sandpaper to remove the rust, rinsed with water. Subsequently, it was immersed into an HCl solution (1 mol·kg^{-1}) for 1~2 minutes, then, take it out and rinsed with water, wipe it to be dry.

Electroplating: the above Cu-Zn galvanic cell was used as power, copper anode was connected with the positive electrode of Cu-Zn galvanic cell, nails used as cathode was connected with the negative electrode of Cu-Zn galvanic cell. In order to avoid contact plating, the nail must be charged when it is immersed into the electrolyte, the device diagram of electroplating was shown in Fig. 2-6. After 10 minutes (when the plating process was on going, please continue to do the next experiment, don't waste time), take out nail to observe whether the copper has been plated.

Take out the copper-plated nail and rinsed with distilled water.

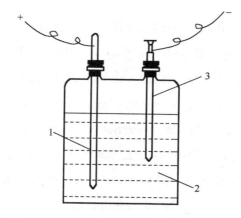

Fig. 2-6 Schematic diagram of electroplating

3. Electrochemical Corrosion of Metal

Adding one zinc particle into a tube with 2 mL of HCl solution (0.1 mol·kg^{-1}), observing its phenomenon. Then a polished copper rod was inserted into this tube and making it be contacted with the zinc particle. Observing the difference before and after the copper rod was inserted. Try to explain this difference (Attention: the zinc particle must be washed and returned to the recovery bottle).

Taking a piece of zinc-plated iron (if there is stain on the surface, please clear it), filing a deep mark in this piece of zinc-plated iron until the iron layer was exposed. Placing it into the drop-board, adding 1~2 drops of HCl solution (1 mol·kg^{-1}) and K$_3$[Fe(CN)$_6$] solution. Observing the experiment phenomenon and answering the question that which metal is eroded? Why?

4. Influence of the Corrosion Inhibitor and Cathode Protection Method

Influence of the corrosion inhibitor: two iron nails which have been polished by sandpapers were placed into two tubes, respectively. 5 drops of urotropine (20%) were added into one of the tubes, and 5 drops of distilled water were added into the other tube. Then, 1~2 mL of HCl solution (1 mol·kg^{-1}) and 1~3 drops of K$_3$[Fe(CN)$_6$] solution (0.1%) were added into each tube. Comparing the difference happened in 2 tubes. Explained it.

Cathode protection method: washing the drop-board, preparing corrosion solution in the groove of drop-board (1 mL of NaCl solution with a concentration of 1 mol·kg^{-1} and 3 drops of $K_3[Fe(CN)_6]$ solution (0.1%)). One piece of filter paper was immersed into the corrosion solution and then was taken out, placing it on the plain of the drop-board. Two iron nails were clamped into the positive and negative electrode of the Cu-Zn galvanic cell, respectively. Placing them in parallel onto the immersed filter paper, waiting for several minutes, observing what will happen and explain it. Adding one drop of phenolphthalein on this filter paper, observing and explaining the phenomenon.

V. Discussion and Exercises

(1) What parts the galvanic cell included? If there is no voltmeter, how the current can be proved to be existed?

(2) We said that the potential of the galvanic just can be determined roughly by a voltmeter, why?

(3) At the experiment of electrochemical erode of zinc-plated iron, why the zinc plate layer was firstly eroded?

(4) At the experiment of the reaction between Zn and HCl solution, why the reaction rate can be accelerated by inserting a copper rod?

Attachment

1. Urotropine

Urotropine, another name is hexamethylenetetramine, which is a heterocyclic organic compound with the formula $(CH_2)_6N_4$ solution. Three-dimensional cage molecules, the structure is shown in Fig. 2-7. This white crystalline compound is highly soluble in water and polar organic solvents. It has a cage-like structure similar to adamantane. It is useful in the synthesis of other chemical compounds, for example, plastics, pharmaceuticals, rubber additives. It sublimes in a vacuum at 280 °C.

It can be used as the corrosion inhibitor is because that it can react with the

H⁺ to form salt in the acid medium. The resulting salt was adsorbed onto the surface of the metal so that the H⁺ in the solution became difficult to react with metal. Therefore, the corrosion rate became slower.

Fig. 2-7　The Structure of urotropine

2. Composition of Plating Solutions

The following compositions were included in the plating solution: $CuSO_4$ solution (10~15 g), $H_2C_2O_4$ solution (60~100 g), ammonia (65~80 mL).

Experiment 5 Inorganic Compound

I. Objects

(1) Understand the synthesis dissociation and transformation of coordination compounds.

(2) Understand the properties of silver halide.

(3) Understand the redox properties of chromium, manganese, iron compounds with different valence.

(4) Understand the effects of medium on the redox properties.

II. Principles

1. The Formation of Coordination Compounds

One of the characteristics of the subgroup elements in periodic table is that they are easy to form coordination compounds. Most of the coordination compounds are comprised of internal and external ion, the internal ion was comprised of the central ion and coordination body. The common coordination ion was as follows: $[Ag(NH_3)_2]^+$, $[Fe(SCN)]^{2+}$(blood red), $[FeF_6]^{3-}$, $[Ag(S_2O_3)_2]^{2-}$ and so on. The properties such as color, solubility and redox and so on of compounds maybe changed and the stability was improved after coordination compounds are formed.

2. Dissociation Equilibrium of Coordination Ion

Coordination compounds are strong electrolytes, they will be dissociated into coordination ion and the simple external ion. For example:

$$[Ag(NH_3)_2]Cl \rightleftharpoons [Ag(NH_3)_2]^+ + Cl^-$$

The coordination ion is stable, it can be partly dissociated in aqueous solution as like weak electrolyte. For example:

$$[Ag(NH_3)_2]^+ \rightleftharpoons Ag^+ + 2NH_3$$

The dissociation equilibrium of coordination ion is also ion equilibrium. So, the equilibrium will shift when the surround changed. For example:

(1) The following equilibrium can shift when the concentration of Ag^+ or NH_3 changed:

$$[Ag(NH_3)_2]^+ \rightleftharpoons Ag^+ + 2NH_3$$

(2) When F^- was added into the solution of $[Fe(SCN)]^{2+}$ coordination ion, the equilibrium will shift to the direction where more stable $[FeF_6]^{3-}$ will be formed:

$$[Fe(SCN)]^{2+} + 6F^- \rightleftharpoons [FeF_6]^{3-} + SCN^-$$

If two or more than two atoms were connected with a central ion to form a ring structure in a coordination body, then, these compounds are also called chelate. Many metals chelate have characteristic colors, and they are difficult to dissolve in water. So, chelates are often used to identify the metal ions in analytical chemistry. For example, the Ni^{2+} ion can be identified by the chelate of dimethylglyoxime due to that the insoluble red chelate precipitation can be formed by the bond of Ni^{2+} and dimethylglyoxime under weak base condition.

$$[Ni(NH_3)_4]^{2+}(aq) + 2 \begin{array}{c} CH_3-C=NOH \\ | \\ CH_3-C=NOH \end{array}(aq) + 2H_2O(l) + 2OH^-(aq) \rightleftharpoons$$

$$\begin{array}{c} O - \cdots H-O \\ \uparrow \quad \quad | \\ CH_3-C=N \quad \quad N=CH_3-C \\ | \quad \quad \searrow Ni \swarrow \quad \quad | \\ CH_3-C=N \quad \quad N=C-CH_3 \\ | \quad \quad | \\ O-H \cdots - O \end{array}(s) + 4NH_3 \cdot H_2O(l)$$

3. The Properties of Silver Halide

About the silver halide, AgCl, AgBr, and AgI precipitation are white, light yellow and yellow, respectively, some precipitation maybe dissolved in $NH_3 \cdot H_2O$ or $Na_2S_2O_3$ solution due to the formation of $[Ag(NH_3)_2]^+$ or $[Ag(S_2O_3)_2]^{3-}$. Such as:

$$AgCl + 2NH_3 \rightleftharpoons [Ag(NH_3)_2]^+ + Cl^-$$

$$AgCl + 2S_2O_3^{2-} \rightleftharpoons [Ag(S_2O_3)_2]^{3-} + Cl^-$$

$$AgBr + 2S_2O_3^{2-} \rightleftharpoons [Ag(S_2O_3)_2]^{3-} + Br^-$$

4. Redoxation of Some Compounds

Many elements in dregion of the periodic table have many valences. Such as: the main valence of manganese element includes +2, +4, +6 and +7, chromium element has +3 and +6 valence. The iron element has +2 and +3 valence. These elements with high valence displayed oxidation, and the elements with low valence possess reduction, the element with intermediate valence displayed both oxidation and reduction.

(1) Both $KMnO_4$ and $K_2Cr_2O_7$ are strong oxidizing agent, their oxidation depend on the medium (acid, neutral or base). For example, the ion reactions between $KMnO_4$ and Na_2SO_3 in different mediums are as following:

$$2MnO_4^- + 5SO_3^{2-} + 6H^+ \rightleftharpoons 2Mn^{2+} + 5SO_4^{2-} + 3H_2O$$

$$2MnO_4^- + 3SO_3^{2-} + H_2O \rightleftharpoons 2MnO_2\downarrow + 3SO_4^{2-} + 2OH^-$$

$$2MnO_4^- + SO_3^{2-} + 2OH^- \rightleftharpoons 2MnO_4^{2-} + SO_4^{2-} + H_2O$$

(2) The highest valence of chromium element is +6, both the CrO_4^{2-} and $Cr_2O_7^{2-}$ can be existed when the pH value was different. They can be transferred mutually in different pH as following:

$$2CrO_4^{2-} + 2H^+ \rightleftharpoons Cr_2O_7^{2-} + H_2O$$

In the acid medium, SO_3^{2-} can be oxidized into SO_4^{2-} by $Cr_2O_7^{2-}$ which has strong oxidation property, the ion reaction was as following:

$$Cr_2O_7^{2-}+3SO_3^{2-}+8H^+ = 2Cr^{3+}+3SO_4^{2-}+4H_2O$$

(3) The redox property of the element with the intermediate valence can be discussed using hydrogen peroxide (H_2O_2) as example. The oxidation number of H_2O_2 is −1, so it characteristic chemical property is oxidation and instability. In certain conditions, it also can displayed reduction. Such as:

$$H_2O_2+2I^-+2H^+ = I_2+2H_2O$$

$$2MnO_4^-+5H_2O_2+6H^+ = 2Mn^{2+}+5O_2\uparrow+8H_2O$$

III. Apparatus and Reagents

1. Apparatus

Apparatus Name	Specification	Number
Test tube	10 mL	8
Centrifuge tube	5 mL	3
Test tube shelf		1
Centrifuge		1

2. Reagents

Reagents Name	Concentration	Reagents Name	Concentration
$AgNO_3$ solution	0.1 mol·L^{-1}	$NH_3·H_2O$ solution	2 mol·L^{-1}
NaOH solution	2 mol·L^{-1}	$FeCl_3$ solution	0.1 mol·L^{-1}
KSCN solution	0.1 mol·L^{-1}	NaF solution	0.1 mol·L^{-1}
$NiSO_4$ solution	0.1 mol·L^{-1}	$NH_3·H_2O$ solution	6 mol·L^{-1}
Dimethylglyoxime		NaCl solution	0.1 mol·L^{-1}
KBr solution	0.1 mol·L^{-1}	KI solution	0.1 mol·L^{-1}
$KMnO_4$ solution	0.01 mol·L^{-1}	H_2SO_4 solution	3 mol·L^{-1}
Distilled water solution		NaOH solution	6 mol·L^{-1}
Na_2SO_3 solution	0.5 mol·L^{-1}	$K_2Cr_2O_7$ solution	0.1 mol·L^{-1}
K_2CrO_4 solution	0.1 mol·L^{-1}	HNO_3 solution	2 mol·L^{-1}
H_2O_2 solution	3%	$Na_2S_2O_3$ solution	0.2 mol·L^{-1}

IV. Procedures

1. Formation and Dissociation of Ag (I) Coordinated Ion

3 drops of $AgNO_3$ solution (0.1 mol·L^{-1}) was added into a test tube, then $NH_3·H_2O$ solution (2 mol·L^{-1}) was added drop by drop. The test tube need to be oscillated fully after each drop $NH_3·H_2O$ solution was added. When the precipitation was disappeared, adding more 1~2 drops of $NH_3·H_2O$ solution (2 mol·L^{-1}). Observing the experimental phenomenon and writing the reaction equations.

2. Formation and Dissociation of Fe (III) Coordinated Ion

2 drops of $FeCl_3$ solution (0.1 mol·L^{-1}) were added into a test tube. This solution was diluted to be colorless. Then 1~2 drops of KSCN solution (0.1 mol·L^{-1}) were added, observing and recording the experimental phenomenon.

Add NaF solution (0.1 mol·L^{-1}) to the above test tube, observe the color change, write the reaction equations and explain this phenomenon.

3. Formation and Color Change of Ni (II) Coordinated Compound

0.5 mL $NiSO_4$ solution (0.1 mol·L^{-1}) was added into a test tube, then 0.5 mL ammonia was added. Observe and record the experimental phenomenon.

Add 1~2 drops of dimethylglyoxime solution into the above solution, observing the red precipitation.

4. Properties of Silver Halide

Preparing 3 test tubes, 2 drops of NaCl solution (0.1 mol·L^{-1}) was added in each tube. Subsequently, 5 drops of $AgNO_3$ solution (0.1 mol·L^{-1}) was added in each tube to form AgCl solution precipitation. Then, 3 three centrifuge tubes were placed into centrifuge sleeves to centrifuge (The operation method of centrifuge was attached as an appendix after this experiment). The upper suspension was pour out, then 5~6 drops of HNO₃ solution (2 mol·L⁻¹), NH₃·H₂O solution (2 mol·L⁻¹) and Na₂S₂O₃ solution (0.2 mol·L⁻¹) were added into the 3 centrifuge tubes, respectively. Observing and recording the experimental phenomenon, then writing the reaction equations.

According the above method, replacing the NaCl solution by KBr solution (0.1 mol·L⁻¹) and KI solution (0.1 mol·L⁻¹), respectively. Completing the similar experiments, observing and recording the experimental phenomenon, then writing the reaction equations.

5. Oxidation Property of KMnO₄

Taking 3 test tubes, 5 drops of KMnO₄ solution (0.01 mol·L⁻¹) were added in each tube. Subsequently, 2 drops of H₂SO₄ solution (3 mol·L⁻¹), distilled water, and NaOH solution (6 mol·L⁻¹) were added into the 3 tubes, respectively. Observing and recording the experimental phenomenon, then writing the reaction equations.

6. Reciprocal Transformation between $K_2Cr_2O_7$ and K_2CrO_4

0.5 mL K₂Cr₂O₇ solution (0.1 mol·L⁻¹) was added into one test tube, 0.5 mL K₂CrO₄ solution (0.1 mol·L⁻¹) was added into another test tube. Observing the color of two solutions and recording it. Then 1 drop of NaOH solution (6 mol·L⁻¹) was added into the first tube and 1 drop of H₂SO₄ solution (3 mol·L⁻¹) was added into the second tube. Observing the change of color and explain why. Writing the reaction equations.

7. The Redox Properties of H_2O_2

10 drops of KI solution (0.1 mol·L⁻¹), 2~3 drops of H₂SO₄ solution and 5~6 drops of H₂O₂ solution (3%) were mixed, observing the phenomenon and writing the reaction equations, pointing out which is the oxidizing agent.

5 drops of KMnO₄ solution (0.01 mol·L⁻¹), 2~3 drops of H₂SO₄ solution and

5~6 drops of H_2O_2 solution (3%) were mixed, observing the phenomenon and writing the reaction equations, pointing out which is the reducing agent.

V. Discussion and Exercises

(1) In the solutions of $AgNO_3$ and $[Ag(NH_3)_2]NO_3$ with the same concentration, whether the concentration of Ag^+ or NO_3^- are the same?

(2) When NaF solution was added into the $[Fe(SCN)]^{2+}$ solution, the color changed, why?

(3) What's the differences of the solubility of AgCl, AgI, AgBr in HNO_3, $NH_3 \cdot H_2O$ and Na_2SO_3 solutions?

(4) Whether the oxidizing property of $KMnO_4$ solution is same in aid, neutral or base medium?

(5) $K_2Cr_2O_7$ solution and K_2CrO_4 solution can transform reciprocally, whether the valence of Cr element has changed?

Attachment

Operation Method of Centrifuge

The centrifuge was used to separate the sediment from the solution. Several points need to be carefully when using it:

(1) Centrifuge tubes must place into the centrifuge sleeves and the position in centrifuge should be symmetric. Otherwise, the machine is easy to be damaged. If the number of tubes needing to centrifuge is one or five, then another tube loading with water at the same volume should be placed to balance.

(2) Turning on the knob gently, making the rotation speed increased gradually. Turning off the knob after 1~2 minutes. Don't start centrifuge rudely! It's dangerous!

Experiment 6 Determination of Dissociation Constant of Bromothymol Blue by Spectrophotometric Method

I. Objects

(1) Understand the basic principle of the determination of dissociation constant of weak electrolyte by spectrophotometric method.

(2) Master the operation of pH meter.

(3) Master the preparation method of solution.

II. Principles

Bromothymol blue, a commonly used acid-base indicator, is a weak electrolyte, which partially dissociates in solution as following:

$$HIn \rightleftharpoons H^+ + In^-$$

The dissociation constant can be expressed as following:

$$K_a^\ominus = \frac{Cr(In^-)Cr(H^+)}{Cr(HIn)} \quad (1)$$

Taking the negative logarithm at both sides of this equation, pH value can be expressed as following:

$$pH = pK_a^\ominus + \lg\frac{Cr(In^-)}{Cr(HIn)} \quad (2)$$

Then, the value of $pK_a^\ominus$ at a certain pH value can be obtained once the value of $\frac{[In^-]}{[HIn]}$ determined. Because both HIn and In$^-$ have adsordance at visible light (Fig. 2-8), the value of $\frac{[In^-]}{[HIn]}$ can be determined by spectrophotometric method.

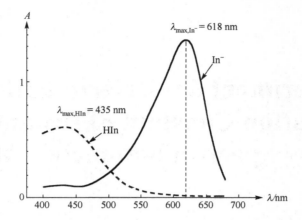

Fig. 2-8　The adsorption spectrum of bromothymol blue

When the pH value of a solution is lower than 4, the bromothymol blue molecules nearly dissociates. Then, the relationship between the absorbance and the concentration of solution as following:

$$A_{\mathrm{HIn}}^{0} = \varepsilon_{\lambda,\mathrm{HIn}} b(\mathrm{HIn}) = \varepsilon_{\lambda,\mathrm{HIn}} b c^{0} \tag{3}$$

Similarly, when the pH value of the solution is higher than 10, almost of the bromothymol blue molecules dissociate into In⁻, then, the relationship between the absorbance and the concentration of solution is as following:

$$A_{\mathrm{In}^{-}}^{0} = \varepsilon_{\lambda,\mathrm{In}^{-}} b(\mathrm{In}^{-}) = \varepsilon_{\lambda,\mathrm{In}^{-}} b c^{0} \tag{4}$$

When the pH value is between 4 and 10, the bromothymol blue solution is partially dissociated, the absorbance can be expressed as following:

$$A_{x} = \varepsilon_{\lambda,\mathrm{HIn}} b(\mathrm{HIn}) + \varepsilon_{\lambda,\mathrm{In}^{-}} b(\mathrm{In}^{-}) \tag{5}$$

According to the material conservation rule, the sum of the equilibrium concentration of HIn and In⁻ equal to the concentration of the original electrolyte c^{0}, namely as following:

$$c^{0} = c(\mathrm{HIn}) + c(\mathrm{In}^{-}) \tag{6}$$

Combining equations (3),(4), (5)and (6), we can obtain an equation as following:

$$\frac{c(\mathrm{In}^{-})}{c(\mathrm{HIn})} = \frac{A_{\mathrm{HIn}}^{0} - A_{x}}{A_{x} - A_{\mathrm{In}^{-}}^{0}} \tag{7}$$

Replacing equation (2) by equation (7), we obtained an equation as following:
$$\mathrm{pH} = \mathrm{p}K_a^\ominus + \lg \frac{A_{\mathrm{HIn}}^0 - A_x}{A_x - A_{\mathrm{In}^-}^0} \tag{8}$$

Where, ε —— Molar absorption coefficient;

b —— The thickness of the quartz curette;

c^0 —— The original concentration of the solution;

A_{HIn}^0 —— The absorbance in strong acid medium, where, bromothymol blue nearly dissociates;

$A_{\mathrm{In}^-}^0$ —— The absorbance in strong base medium where bromothymol blue dissociates completely;

A_x —— The absorbance in pH value between 4 and 10, which is measured exactly by a pH meter.

Therefore, $\mathrm{p}K_a^\ominus$ can be determined through equation (8). However, in order to eliminate the error, the value of $\mathrm{p}K_a^\ominus$ was generally determined by making the curve of pH and $\lg \frac{A_{\mathrm{HIn}}^0 - A_x}{A_x - A_{\mathrm{In}^-}^0}$. As shown in Fig. 2-9, the value of $\mathrm{p}K_a^\ominus$ can be obtained by the intercept on vertical axis.

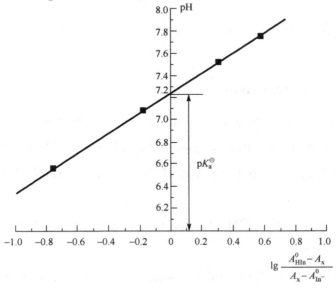

Fig. 2-9 Determination of $\mathrm{p}K_a^\ominus$ by the curve of pH and $\lg \frac{A_{\mathrm{HIn}}^0 - A_x}{A_x - A_{\mathrm{In}^-}^0}$

III. Apparatus and Reagents

1. Apparatus

Apparatus Name	Specification	Number
722S-Spectrophotometer		1
pHS-3C pH meter		1
Colorimetric tube	25 mL	71
Pipette	1 mL	1

2. Reagents

Reagents Name	Specification
NaH_2PO_4 solution	$0.2 \ mol \cdot L^{-1}$
K_2HPO_4 solution	$0.2 \ mol \cdot L^{-1}$
HCl solution	$6 \ mol \cdot L^{-1}$
NaOH solution	$4 \ mol \cdot L^{-1}$
Buffer solution for pH meter calibration	pH=6.8
Buffer solution pH meter calibration	0.1%

IV. Procedures

1. Preparation of Solution

(1) Prepare 7 colorimetric tubes, number them from 1 to 7.

(2) Pipette 1.00 mL of bromothymol blue solution (0.1%(w/v)) to each tube.

(3) Add 4 drops of HCl solution ($6 \ mol \cdot L^{-1}$) to the No.1 tube to make a strong acid solution.

(4) Add 5 drops of NaOH solution ($4 \ mol \cdot L^{-1}$) to the No.7 tube to make a strong baisc solution.

(5) For the No.2~6 tubes, add phosphate solutions according to the volume shown in Tab.2-4 to make a series of solution at pH of 4 to10.

Tab. 2-4 Reagents dosage

Room temperature:_____°C   Wavelength λ:_____nm   Date:_____

No.	Bromothymol Blue Solution /mL	NaH$_2$PO$_4$ Solution /mL	K$_2$HPO$_4$ Solution /mL	Other Reagents	pH	A	$\dfrac{c(\text{In}^-)}{c(\text{HIn})} = \dfrac{A^0_{\text{HIn}} - A_x}{A_x - A^0_{\text{In}^-}}$	p$K_a^\ominus$
1	1.00	0	0	4 drops of HCl Solution		$A^0_{\text{HIn}} = 0$		
2	1.00	2.5	0.5					
3	1.00	5.0	2.5					
4	1.00	2.5	5.0					
5	1.00	0.5	2.5					
6	1.00	0.5	5.0					
7	1.00	0	0	5 drops of NaOH Solution		$A^0_{\text{In}^-} = ?$		

2. Determination of the Absorbance of Bromothymol Blue Solutions by Spectrophotometer

The absorbance of No.2~7 solutions at wavelength of 618 nm are measured by a spectrophotometer using solution No.1 as the reference solution (Please find the working principle of the 722S-type spectrophotometer in appendix 1 followed this experiment.)

(1) Turn on the power, open the lid of the colorimetric box to switch off the light, and warm up the instrument for 20 minutes.

(2) Set the wavelength at 618 nm by turning the white knob on upper surface.

(3) Prepare the measured solutions as follows:

a) Clean the cuvette with distilled water for three times.

b) Rinse the cuvette with measured solutions for three times (Note: add one third volume of measured solution to cuvette to rinse, then trasfer the solution to waste.).

c) Fill the cuvette with measured solutions at volume of 4/5.

Notes:

Make sure only hold the frosted glass surfaces of the cuvette. Do not touch the optical glass of the cuvette.

Wash the cuvette with tap water and distilled water thoughly before rinse with testing solutions to avoid dilution or containmination of the testing solution when add to the cuvette.

Make sure wipe outside of the cuvette before measurement if there are water or solution on the outside walls of the cuvette.

(4) Place cuvettes containing the reference solution (solution NO.1) and the testing solution (solutions NO.2~7) in the colorimetric shelf. Make sure place the reference solution on the first lattice and the testing solutions on other lattices in queue.

(5) Calibration of "0%". Open the lid of the colorimetric box, to cut off the light . Press the "0%" button continuously till the screen displays "0.0000".

(6) Calibration of "100%". Set the reference solution on the optical pathway. Then, close the lid of the colorimetric box and press the button of "100%" continuously till the screen displays "100%".

(7) Measurement. Press the "mode" button and select the "absorbency" mode , the screen should display "0.0000".Then set the solution No.2 on the optical pathway and record the value of absorbency. Subsequently, record the absorbency of other solutions under the same operation.

(8) Finally, turn off the power, take out the cuvettes to wash and dry before turn back to the colorimetric shelf.

3. Determination of the pH Value of Bromothymol Blue Solutions by a pH Meter

Notes:

Do not keep the electrode in air without solution or rinse in a testing solution for a long time. Otherwise, immerse the electrode into distilled water for 8 hours before calibration or measurement.

Make sure remove the protective sleeve at the bottom of the electrode before measurement.

Plug the power and turn on the switch to preheate the pH meter for 10 minutes before calibration or measurement. (Please find the operation chart of the pHS-3C pH meter in Fig.2-10 and the working principle of pHS-3C pH meter at the second appendix followed this experiment).

(1) Calibrations of the pH meter.

a) Accuracy calibration.
- Insert the electrode of the pH meter into a buffer solution at pH of 6.8.
- press the button of "accuracy calibiration", adjust the displayed pH value to 6.80.
- press the confirm button.

b) Slope coefficient calibiration.
- Insert the electrode of the pH meter into a buffer solution at pH of 4.0.
- Press the button of "Slope coefficient calibration".
- Adjust the displayed pH value to 4.00.
- Press the confirm button.

c) Temperature calibration.
- Press the button of "temperature calibration".
- Adjust the displayed temperature to room temperature.
- Press the confirm button.

Note:

Do not touch these parameter setting buttons anymore after calibrations!

(2) Measurement of pH value of Solutions No. 2~6.

a) Wash the electrode with distilled water.

b) Wipe the dry the electrode before measurement (Note: dry the electrode carefully and gently as the sensitive glass bulb in the front of the electrode was surper fragile!).

c) Insert the electrode to the testing solution and keep for 10 to 30 seconds. (Note: remember only measure the solutions from No.2 to No.6. Do not measure solution No.1 and No.7).

d) Record the pH value shown in screen till the value is stable.

e) Wash the electrode with distilled water and wipe to dry, then measure the next sample.

Notes:

Before and after each measurement, remember to wash the electrode with distilled water and wipe to dry.

When finish all of the measurements, wash the electrode with distilled water and dry before rinse the electrode in distilled water, wear the protective sleeve to the electrode and turn off the power.

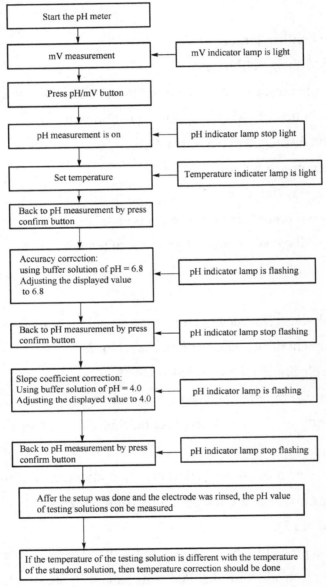

Fig. 2-10 The operation chart of the pHS-3C pH meter

V. Data processing

Mak the plot of pH and $\lg \dfrac{A^0_{HIn} - A_x}{A_x - A^0_{In^-}}$, to obtain a straight line, and the intercept is the value of $pK_a^\ominus$, calculate the relation error finally.

VI. Discussion and Exercises

(1) Describe the principle of the determination of the dissociation constant of bromothymol blue solution.

(2) Whether the dissociation constant is the same for the solutions at different pH?

(3) What should pay attention to when operate a pH meter?

(4) What should pay attention to when operate a spectrophotometer?

Attachment

1. 722S Spectrophotometer

Spectrophotometer is an instrument based on the selective absorption property of matters to monochrome light to determine the substance content.

When a beam of monochromatic light at a certain wavelength goes through a colored solution, a portion of light is absorbed while a portion of light passes through. Define the intensity of incident light as I_0, the intensity of transmission light as I_t, then the I_t / I_0 is regarded as light transmittance, expressed by T as following:

$$T = I_t / I_0$$

The degree of light absorbtion of a colored solution can be expressed by the absorbency A as following:

$$A = \lg(I_0 / I_t)$$

The relations between the absorbency A and the transmittance T can be expressed as following:

$$A = -\lg T$$

Experiment results proved that the absorbency of a solution depend on the concentration, and thickness of the solution and the wavelength of the incident light and so on. If the wavelength of the incident light was kept constant, the absorbency only depend on the concentration and the thickness of the solution. Namely, Lambert-Bill's law as following:

$$A = \varepsilon bc$$

Where, ε —— Molar absorption coefficient ($L \cdot mol^{-1} \cdot cm^{-1}$), it depends on the property of the incident light, the temperature and so on;

b —— The thickness of the solution (cm);

c —— The concentration of the solution ($mol \cdot l^{-1}$).

When the wavelength of the incident light was kept constant, ε is a characteristic constant of the colored substance in solution. Based on Lambert-Bill's law, when the thickness of the solution (b) keeps constant, the absorbency A linears to the concentration of the solution (c). This is the basic principle to determine the content of substances by using a spectrophotometer.

The schematic diagram of the structure of the 722S spectrophotometer is shown in Fig. 2-11.

2. Leici pHS-3C pH Meter

The pH meter comprised of an electrode and a display window is an instrument used to measure both the pH value and the potential of a solution. Actually, the determination of pH value is based on potential measurement. After the potential is measured, pH value can be calculated by a program and directly read in the displayed window.

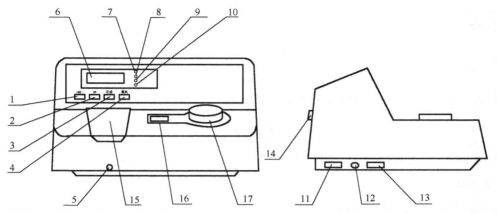

1—100%T; 2—0%T; 3—Extended function button; 4—Display scale; 5—Pull rod of the sample shelf; 6—Display window; 7—Transmission ratio; 8—Absorbency; 9—Concentration factor; 10—Directly reading concentration; 11—Power socket; 12—Fuse; 13—Switch; 14—Serial interface; 15—Sample room; 16—Display window of wavelength; 17—Turning knob of wavelength.

Fig. 2-11 The schematic diagram of the structure of the 722S spectrophotometer

To determine the pH value, the composite electrode is shown in Fig. 2-12 is inserted into the testing solution, a galvanic cell will be formed in the solution as following:

Internal reference electrode	Internal reference solution	Electrode bulb	Testing solution	External reference solution	External reference electrode
$(-)E_{IR}$	E_{IB}	E_{EB}	E_{ET}		$E_{ER}(+)$

Where, E_{IR} —— The electric potential difference between the internal reference electrode and the internal reference solution;

E_{IB} —— The electric potential difference between the internal reference solution and the internal wall of the glass electrode bulb;

E_{EB} —— The electric potential difference between the external wall of the glass electrode bulb and the testing solution;

E_{ET} —— The interfacial electric potential between the external reference solution and the testing solution;

E_{ER} —— The electric potential difference between the external reference electrode and the external reference solution.

The electrode potential of the galvanic cell equal to the sum of every sub-potential, namely as following:

$$E = -E_{IR} - E_{IB} + E_{EB} + E_{ET} + E_{ER}$$

Where,

$$E_{EB} = E^- - \frac{2.303RT}{E}\text{pH}$$

We also can define obsorbency A as following:

$$A = -E_{IR} - E_{IB} + E_{EB} + E_{ET} + E^-$$

When A is a constant under a certain condition:

$$E = A - \frac{2.303RT}{E}\text{pH}$$

Based on the above equation, the electrode potential E is linear to the pH value of the testing solution, then the slope coefficient is $-\frac{2.303RT}{E}$. The constant A will depend on the sub-electrode and the test conditions. So in order to obtain the accurate pH value of different testing solutions, it is necessary to calibrate the pH meter using a standard buffer solution.

The schematic diagram of the structure of the pHS-3C pH meter is shown in Fig. 2-13

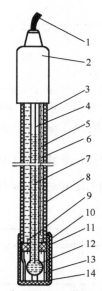

1—Electrode conductance; 2—Electrode cap; 3—Liquid adding hole; 4—Internal reference electrode; 5—External reference electrode; 6—Electrode rod; 7—Internal reference solution; 8—External reference solution; 9—Interdace; 10—Sealing ring; 11—Silicone ring; 12—Electrode bulb; 13—Bulb shield; 14—Sheath.

Fig. 2-12 The composite electrode

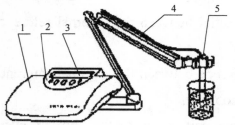 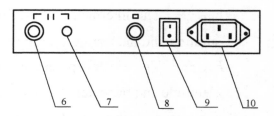

1—Shell of pH meter; 2—Keyboard; 3—Display window; 4—Electrode shelf; 5—Electrode; 6—Electrode socket; 7—Inlet of reference electrode; 8—Fuse; 9—Power switch; 10—Power socket.

Fig. 2-13 The schematic diagram of the structure of the pHS-3C pH meter

Experiment 7 Synthesis of Potassium Trioxalatoferrate (III)

I. Objects

(1) Learn principles and methods of preparation of potassium trioxalatoferrate (III);

(2) Learn basic operations of inorganic synthesis through the preparation of coordination compounds;

(3) Understand the relationship between crystal shape and configuration of coordination compounds.

II. Principles

Potassium trioxalatoferrate (III), a green monoclinic crystal, is soluble in water but insoluble in ethanol, which readily decomposes when exposed to light. It is a main starting material for preparation of active loading iron catalyst and a catalyst for organic reaction.

In this experiment, potassium trioxalatoferrate (III) is prepared through the reaction of ferric chloride and potassium oxalate in solution at 80℃, and the crystal of potassium trioxalatoferrate (III) coordination compound crystallizes in the presence of over-saturated K^+ after the solution cooled to room temperature.

The reaction can be expressed as an ionic equation as following:

$$Fe^{3+}_{(aq)} + 3C_2O_4^{2-}{}_{(aq)} \rightleftharpoons [Fe(C_2O_4)_3]^{3-}_{(aq)} \tag{1}$$

$$[Fe(C_2O_4)_3]^{3-}_{(aq)} + 3K^+_{(aq)} \rightleftharpoons K_3[Fe(C_2O_4)_3]_{(s)} \tag{2}$$

III. Apparatus and Reagents

1. Apparatus

Apparatus Name	Specification	Unit	Quantity
Microscopy (Shared)		Set	4
Decompress suction device (Shared)		Set	6
Water bath		Set	1
Beaker	50 mL	Package	3
Measuring cup	10 mL	Package	1
Suction funnel	Matched with suction device	Package	1
Electronic balance (Shared)	1000 g	Set	3
Scissors (Shared)		Piece	3
Filter paper	Matched with suction funnel	Piece	2
Plastic wash bottle, glass rod, rubber dropper, slides, etc.			One for each group

2. Reagents

Reagents Name	Specification	Chemical Name	Specification
Potassium oxalate	Solid (A.P)	$FeCl_3 \cdot 6H_2O$	Solid(A.P)
Distilled water		Ethanol	

IV. Procedures

(1) Weight out 4.0 g of potassium oxalate in a 50 mL beaker and dissolve completely with 80 mL of distilled water in a 80℃ water bath.

(2) Prepare concentrated solution of ferric chloride using 1.9 g of ferric chloride ($FeCl_3 \cdot 6H_2O$) and 3 mL of distilled water (Note: Do not heat to avoid decomposition). Dropwise this solution to warm potassium oxalate solution with stiring ustilzied glass bar. Observe carefully and record the changes in solution.

(3) Let the reaction stay in water bath for 20~30 minutes. Then stop string and transfer to room temperature to chill it naturally. When crystal started to precipitate in the beaker, pipette 1 or 2 drops of solution by a rubber dropper and drop to a clean slide. Sway the slide gently in the horizontal direction to title the solution in the middle of slide before observe under a microscopy.

(4) Vacuum filter after the mixture of products chilled to room temperature and crystals precipitated almost completely. After filtration, wash the crystals twice with 2 mL of distilled water and twice with 2 mL of ethanol till dry completely.

(5) Weight out the total crystals prepared and calculate the yield. Finally, check by instructors.

V. Data Recording and Processing

1. Reactant

Reactant	$K_2C_2O_4 \cdot H_2O$	$FeCl_3 \cdot 6H_2O$
Specification		
Mass/g		
Amount of substance/mol		

2. Product

Coordination Compound	Crystal Color	Crystal Appearance	Coordinate Number	Yield/g
$K_3[Fe(C_2O_4)_3] \cdot 3H_2O$				

3. Yield

Theoretical product weight (crystal): $W_{theoretical} = $ _____.

Actual product weight (crystal): $W_{actual} = $ _____.

Yield: $\dfrac{W_{actual}}{W_{theoretical}} \times 100\% = $ _____.

VI. Discussion and Exercises

(1) How to define coordinate compounds and ligands? What is the specification of coordinate compounds and ligands compared to general ionic compounds?

(2) How many methods can be used for crystallization? What are the benefits and overcomes for crystals obtained?

(3) Try to design other strategies to prepare potassium trioxalatoferrate (III).

Experiment 8 Determination of Calcium and Magnesium Ion in Water by Complexometric Titrations with EDTA

I. Objects

(1) Learn how to characterize water hardness.

(2) Learn basic principle and operations of determining calcium and magnesium ion contents in water by complexometric titrations with EDTA.

(3) Master condition control in complexometric titrations using calconcarboxylic acid and Eriochrome Black T as visal end-point indicators.

II. Principles

Throughout history, the quality and quantity of water available to humans have been vital factors in determining their welfare. In particular, the acceptable quality is determined by its intended usage. For example, water with a medium salt content could be used for irrigating crops, but not for drinking by humans. Hence, the control of water supplies is a very important in our society.

1. Water Hardness

Water hardness is one of factors to characterize water qulity, which is defined by the concentration of multivalent cations in the water.

Calcium, magnesium and sometimes with Fe^{2+}, Al^{3+}, Mn^{2+}, Sn^{2+}, and Zn^{2+} are common cations found in most fresh-water systems. These ions, generally present as bicarbonates or sulfates, account for water hardness producing an insoluble "curd" by reaction with soap. Total hardness in water is composted by temporary hardness and permanent hardness. Temporary hardness is mainly caused by the presence of dissolved bicarbonate minerals (calcium bicarbonate and magnesium bicarbonate), which can be reduced by boiling water and promotes the formation of carbonate from the bicarbonate and precipitates calcium carbonate out of solution. Permanent hardness is mainly caused by sulphate and chloride compounds which can not be removed by boiling, such as $CaSO_4$, $MgSO_4$, $CaCl_2$, $MgCl_2$, $Ca(NO_3)_2$, and $Mg(NO_3)_2$, and so on.

In China, Water hardness is usually expressed as the milligrams of CaO or $CaCO_3$ equivalent to the total amount of Ca^{2+} and Mg^{2+} present in one liter.

2. Determination of Water Hardness by Complexometric Titration

One widely used method for the determination of water hardness is their complexometric titration with EDTA (Ethylene-Diamino Tetraacetic Acid), using Eriochrome Black T (EBT) or calmagite as visual end-point indicators.

In details, EDTA is used as the titrant that complexes Ca^{2+} and Mg^{2+} ions. EDTA is widely used in analytical chemistry, as it forms strong one-to-one complexes with most metal ions. The neutral acid is tetraprotic and is often denoted as H_4Y (Fig. 2-14).

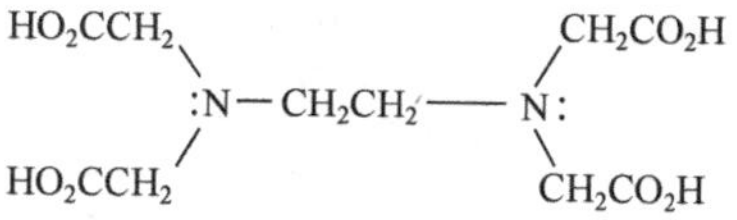

Fig. 2-14 H_4Y: Ethylene Diamino Tetraacetic Acid (EDTA)

At pH 10, a portion of EDTA is in the HY^{3-} form while EDTA is present in its completely deportonated form Y^{4-} when pH values greater than 12. For Ca^{2+} and Mg^{2+}, the complexation reactions with EDTA are as following:

$$Ca^{2+} + HY^{3-} \longrightarrow CaHY^{2-} + H^+ \tag{1}$$

$$Mg^{2+} + HY^{3-} \longrightarrow MgHY^{2-} + H^+ \tag{2}$$

$$Ca^{2+} + Y^{4-} \longrightarrow CaY^{2-} \tag{3}$$

$$Mg^{2+} + Y^{4-} \longrightarrow MgY^{2-} \tag{4}$$

Although neither the EDTA titrant nor its calcium and magnesium complexes are colored, the end point of the titration can be visually detected by adding a metallochromic indicator to the water sample. A color change occurs when the indicator goes from its metal ion–bound form to the unbound form. This color change signals the end point as it takes place when EDTA removes the Mg^{2+} or Ca^{2+} ions bounding to the indicator after complexes all of free Ca^{2+} and Mg^{2+} ions. This removal is possible because the complexaion affinity of EDTA to Mg^{2+} or Ca^{2+} ion is stronger than the indicator. Both EBT and calmagite (H_3In) are often used in water hardness determination. The color change observed at the end point of the titration corresponds to the following reactions:

At pH 10,

$$MgIn^- + HY^{3-} \longrightarrow MgY^{2-} + HIn^{2-} \quad (5)$$

Reagent: Metal ion-indicator-complex   Titrant   Metal ion-EDTA-complex   Free indicator

Color: Wine-red   Colorless   Colorless   Blue

At pH 12,

$$CaEBT^- + HY^{3-} \longrightarrow CaY^{2-} + HEBT^{2-} \quad (6)$$

Reagent: Metal ion-indicator-complex   Titrant   Metal ion-EDTA-complex   Free indicator

Color: Wine-red   Colorless   Colorless   Blue

The precipitation of Mg^{2+} as $Mg(OH)_2$ precludes the titration at pH 12, so the water hardness calculated is attributed only by Ca^{2+}. Hence, the total hardness of water and calcium hardness are calculated as following:

$$\text{Total hardness of water (mg/L)} = \frac{c_{EDTA} V_{EDTA} M_{CaO}}{V_{water}} \times 1000 \quad \text{(Titrate at pH 10)} \quad (7)$$

$$\text{Calcium hardness of water (mg/L)} = \frac{c_{EDTA} V_{EDTA} M_{CaO}}{V_{water}} \times 1000 \quad \text{(Titrate at pH 12)} \quad (8)$$

III. Apparatus and Reagents

1. Apparatus

Apparatus Name	Specification	Unit and Quantity
Beaker	100 mL	1
Conical flask	250 mL	1
Cylinder 1	50 mL	1
Cylinder 2	10 mL	2
Suction pipette	50 mL	1
Rubber/Suction/Bulb		1
Burette	50 mL	1

2. Reagents

Reagents Name	Specification	Unit and Quantity
Running water		
Triethanolamine		
EDTA standard solution	0.01 mol/L	
NH_3-NH_4Cl buffer solution	pH = 10	
Hydroxylamine hydrochloride	1% (W/V)	
Sodium hydroxide	10% (W/V)	
Calconcarboxylic acid		
Eriochrome Black T		

IV. Procedures

1. Determination of Total Hardness of Water

(1) Fill EDTA standard solution to a burette and record the value of solution (V_o).

(2) Pipette 50.00 mL of running water using a 50 mL suction pipette to a 250 mL conical flask.

(3) Add 1 mL of triethanolamine, 1 mL of hydroxylammonium chloride to

the water (1 mL of solution is roughly 16 to 20 drops with glass pipette) and shake to mix well.

(4) Keep in room temperature for 2~3 minutes.

(5) Add 10 mL of buffer solution, and 1 drop of EBT, and titrate with EDTA standard solution immediately till the solution changed from purple red to blue (Note: titrate slowly and shake well after each drop of titration).

(6) Record the final value of EDTA solution (V_f).

(7) Repeat twice of step (1) to step (6).

(8) Calculate the total hardness in water, expressed as milligrams CaO per liter of water.

2. Determination of Ca^{2+} Hardness in Water

(1) Fill EDTA standard solution to a burette and record the value of solution (V_o).

(2) Pipette 50.00 mL of running water using a 50 mL suction pipette to a 250 mL conical flask.

(3) Add 5 mL of sodium hydroxide at concentration of 10% (W/V) to the water and shake to mix well (Note: adjust the pH to 12 and form the precipitation of $Mg(OH)_2$ to mask Mg^{2+}).

(4) Add 1 drop of calconcarboxylic acid and shake to mix well (The solution should be wine-red).

(5) Titrate with EDTA standard solution immediately till the solution changed from pink to blue (Note: titrate slowly and shake well after each drop of titration).

(6) Record the final value of EDTA solution (V_f).

(7) Repeat twice of step (1) to step (6).

(8) Calculate the Ca^{2+} hardness in water, expressed as milligrams CaO per liter of water.

V. Data recording and Analyzing

(1) Determination of total hardness in water (Tab. 2-5).

Tab. 2-5 Determination of total hardness in water

Number	1	2	3
$c(EDTA)/mol \cdot L^{-1}$		0.01	
V_{sample}/mL		50.00	
$V_0(EDTA)/mL$			
$V_f(EDTA)/mL$			
$V_1(EDTA)/mL$	$V_1'=$	$V_1''=$	$V_1'''=$
$V_2(EDTA)/mL$			
Total hardness/mg·L^{-1}			

(2) Determination of Ca^{2+} hardness in water (Tab. 2-6).

Tab. 2-6 Determination of Ca^{2+} hardness in water

Number	1	2	3
$c(EDTA)/mol \cdot L^{-1}$		0.01	
V_{sample}/mL		50.00	
$V_0(EDTA)/mL$			
$V_f(EDTA)/mL$			
$V_1(EDTA)/mL$	$V_1'=$	$V_1''=$	$V_1'''=$
$V_2(EDTA)/mL$			
Ca^{2+} hardness/mg·L^{-1}			

VI. Discussion and Exercises

(1) How to define the water hardness? How many expresses of the water hardness?

(2) What should pay attention to when determine the water hardness using EDTA?

Experiment 9 Determination of Mo (VI) Content by Visual Catalytic Kinetics Method

I. Objects

(1) Know the approach of determining traced Mo (VI) through Landolt affection.

(2) Practice the operation of suction pipette.

(3) Learn how to design an experiment based on your knowledge.

II. Principles

Molybdenum has been widely used in metallurgical industry as an additive in alloy steel production to improve elevated temperature strength, wear-resisting property and erosion of metallic materials. It can also be used as heating material and structural material of high-temperature electric resistance and nuclear reactor as well as electrod, grid, semiconductor and electric light source of electron tubes. In chemical industry, molybdenum compounds have mainly be used as lubricants, catalysts and paints. Besides, Mo (VI) is an essential chemical element for growing development and metabolize of organisms. Meanwhile, it has getting more attention to its influence on environment for human survival. Hence, it is important to set a method for traced Mo (VI) determination.

In this experiment, a visual catalytic kinetics method based on Landolt affection is investigated.

In a acidic solution, there is a reaction between $KBrO_3$ solution and KI solution:

$$6I^- + BrO_3^- + 6H^+ \rightleftharpoons 3I_2 + Br^- + 3H_2O \tag{1}$$

The reaction rate could represent as following:

$$v = kc(BrO_3^-) \cdot c(I^-) \cdot c^2(H^+)$$

In which,

v —— The reaction rate;

$c(BrO_3^-)$, $c(I^-)$ and $c(H^+)$ —— The initial concentration of BrO_3^-, I^- and H^+ ($mol \cdot L^{-1}$);

k —— Rate constant.

Landolt affection would happen in the presence of $Na_2S_2O_3$ and starch-iodide indicator, and the reaction is as following:

$$2S_2O_3^{2-} + I_2 \rightleftharpoons S_4O_6^{2-} + 2I^- \tag{2}$$

The reaction (2) is much faster than reaction (1). Hence, the I_2 produced by reaction (1) would react immediately with $S_2O_3^{2-}$ to form $S_4O_6^{2-}$ and I^- which are colorless. The I_2 produced by reaction (1) would react immediately with starch-iodide indicator to show blue in the solution once $Na_2S_2O_3$ reacts completely. So the induction time is from the starting of reaction to the time when blue color shows first in solution.

Mo^{6+} catalyze reaction (1) significantly and its concentration related to $1/t$ linearly which can be described as following:

$$\frac{1}{t} = a + bc(Mo(VI)) \tag{3}$$

The induction time t_0 is defined as the reaction time without catalyst. In general, t_0, a and b are constant. So, the concentration of Mo (VI) can be calculated from equation (3) once the induction time is measured.

III. Apparatus and Reagents

1. Apparatus

Apparatus Name	Item Mode	Specification	Quantity
Electric magnetic stirrer	85-1		1

Continued

Apparatus Name	Item Mode	Specification	Quantity
Magnetic stirrer bar			1
Stopwatch			
Water bath	HH-2		3
Beaker		100 mL	1
Suction pipette 1		5 mL	1
Suction pipette 2		1 mL	1
Rubber suction bulb			1
Glass thermometer		0~100℃	1
Test tubes		50 mL	8
Measuring flask		100 mL	1
Measuring cylinder		10 mL	5
Glass rod		150~200mm	1

2. Reagents

Reagents Name	Concentration (mol·L^{-1})
KI solution	0.01
KBrO$_3$ solution	0.04
Na$_2$S$_2$O$_3$ solution	0.001
HCl solution	0.1
Standard solution with Mo (VI)	0.2 mg/mL of Mo (VI)
Unknown sample with Mo (VI)	0.2 mg/mL of Mo (VI)
Starch-iodide indicator solution	2%
Doistilled water solution (D-H$_2$O)	

IV. Procedures

1. Influence of Temperature on Induction Time

Reagent amounts for determining induction time influenced by temperature is shown in Tab. 2-7.

(1) Prepare two solutions in test tubes at room temperature. The first one is a mixture of 5 mL of HCl solution, 5 mL of KBrO, 35 mL of D-H$_2$O solution and 2 drops of starch-iodide indicator solution. The second is a mixture of 5 mL of KI solution and 5 mL of Na$_2$S$_2$O$_3$ solution. Place the second solution into the first

tube and stir with a glass rod. Set starting time by stopwatch when two solutions mixed. Record ending time once blue color is observed in the react solution, which is the induction time t_0 (it is the control value).

(2) At room temperature, pipette 1.0 mL of standard solution with Mo (VI) using a suction pipette to a test tube. Then, add 5 mL of HCl solution, 5 mL of KBrO$_3$ solution, 4.8 mL of D-H$_2$O solution and 2 drops of starch-iodide indicator solution. Shake to mix well. In another test tube, add 5 mL of KI solution and 5 mL of Na$_2$S$_2$O$_3$ solution, and shake to mix well. Then, transfer the mixture in the second tube to the first tube immediately and stir with a glass rod. Set starting time by stopwatch when two solutions mixed. Record ending time once blue color is observed in the react solution, which is the induction time (t).

(3) Repeat step (1) and step (2) at different temperatures (try the temperature 5℃, 10℃ and 15℃ higher than room temperature). Record the induction times. Discuss how the temperature affects induction time.

Tab. 2-7 Reagent amounts for determining induction time influenced by temperature

Experiment Number (No.)		Group 1 Room Temperature (r.t.)		Group 2 r.t.+5℃		Group 3 r.t.+10℃		Group 4 r.t.+15℃	
		1-1	1-2	2-1	2-2	3-1	3-2	4-1	4-2
Test tube 1 /mL	Mo (VI) solution (0.2 mg/ mL)	0	1	0	1	0	1	0	1
	HCl solution (0.1/mol·L^{-1})	5	5	5	5	5	5	5	5
	KBrO$_3$ solution (0.04/mol·L^{-1})	5	5	5	5	5	5	5	5
	D-H$_2$O solution	5	5	5	5	5	5	5	5
	Starch-iodide indicator (2%)	2 drops	2 drops	2 drops	2 drops	2 drops	2 drops	2 drops	2 drops
Test tube 2 /mL	KI solution (0.01/mol·L^{-1})	5	5	5	5	5	5	5	5
	Na$_2$S$_2$O$_3$ solution (0.001/mol·L^{-1})	5	5	5	5	5	5	5	5
Concentration of Mo (VI) /(μg/ mL)									
Induction time (t/s)		$t_0=$	$t=$	$t_0=$	$t=$	$t_0=$	$t=$	$t_0=$	$t=$
$\frac{1}{t}-\frac{1}{t_0}$/s^{-1}									

2. Working Curve of Mo (VI) Concentration Determination

Prepare two solutions in two beakers at room temperature. Pipette standard solution with Mo (VI) using a suction pipette to a 100 mL beaker. Then add HCl solution, KBrO₃ solution, D-H₂O solution and 3 drops of starch-iodide indicator solution. Place the beaker on a electric magnetic stirrer to mix the solution well. Mix KI solution and Na₂S₂O₃ solution in another 100 mL beaker and shake to mix well before add to the first beaker with stirring immediately. Set starting time by stopwatch when two solutions mixed. Record ending time once blue color is observed in the react solution, which is the induction time (t). The reagent amounts is shown in Tab. 2-8.

Tab. 2-8 Reagent amounts for determining working curve

		Experiment Number (No.)	1	2	3	4	5
Reagent amount /mL	Beaker 1	Mo (VI) solution (0.2 mg/mL)	0	0.5	1	2	4
		HCl solution (0.1/mol·L^{-1})	10	10	10	10	10
		KBrO₃ solution (0.04/mol·L^{-1})	10	10	10	10	10
		D-H₂O solution	10	8.5	9	8	6
		Starch-iodide indicator solution (2%)	3 drops	3 drops	3 drops	3 drops	3 drops
	Beaker 2	KI solution (0.01/mol·L^{-1})	10	10	10	10	10
		Na₂S₂O₃ solution (0.001/mol·L^{-1})	10	10	10	10	10
	Concentration of Mo (VI)/(μg/mL)						
	Induction time (t/s)						
	$\frac{1}{t}$/s^{-1}						

Plot $1/t$ (unit is s^{-1}, vertical coordinates) verse concentration of Mo (VI) (unit is μg/mL, horizontal coordinates) for each group to give the working curve and linear regression equation is then given by the software (Origin 8.0).

3. Determination of Mo (VI) Concentration in Unknown Sample

Prepare two solutions in two beakers at room temperature. Pipette 1 mL of unknown sample with Mo (VI) using a suction pipette to a 100 mL beaker. Then add 10 mL of HCl solution, 10 mL of $KBrO_3$ solution, 9 mL of D-H_2O solution and 3 drops of starch-iodide indicator solution. Place the beaker on a electric magnetic stirrer to mix the solution well. Mix 10 mL of KI solution and 10 mL of $Na_2S_2O_3$ solution in another 100 mL beaker and shake to mix well before add to the first beaker with stirring. Set starting time by stopwatch when two solutions mixed. Record ending time once blue color is observed in the react solution, which is the induction time (t). Repeat once and calculate the Mo (VI) concentration in unknown sample through the linear regression equation given by last section. Finally, calculated the mean value of Mo (VI) concentration in unknown sample and analyze the measuring error.

V. Discussion and Exercises

(1) Why the reaction temperature should be controlled in this experiment? How to choose reaction temperature?

(2) What factors mainly influence the measured induction time? How to minimize the measuring error of induction time?

(3) Try to design an experiment to understand whether $Na_2S_2O_3$ can be replaced by other reagents such as thiocarbamide, hydroxylamine hydrochloride and ascorbic acid.

Experiment 10 Separation of Coordination Compounds by Ion Exchange Resin

I. Objects

(1) Know basic principle of ion exchange separation.

(2) Learn operations of ion exchange separation.

(3) Learn operations of UV-Vis spectrometer.

II. Principles

Ion exchange separation is one of the most common separation methods, espeically for the separations of those samples who have similar properties or element with low content.

Ion exchange resin is a polymer mixture with high molecular weight matrix polymerized by monomers such as styrene and divinylbenzene. The exchangeable groups introduced by chemical strategies endow the resin activity. It can be cation exchange resin and anion exchange resin according to the active groups while it can be strong base, strong acid, weak base and weak acid resin and so on.

The structure formula of cation exchange resin used in this experiment can be expressed as $RSO_3^-H^+$. When a solution containing cation M^+ pass through this resin, M^+ will attract RSO_3^- and substitute H^+ in the resin. The degree of substitution depends on the properties and concentration of M^+. The process can be expressed as following:

$$RSO_3^-H^+ + M^+ \Longleftrightarrow RSO_3^-M^+ + H^+ \tag{1}$$

Each M^+ has its own equilibrium constant while the equilibrium point depends on the relative concentration of M^+ and H^+ in solution. M^+ combines RSO_3^- in the highest degree when concentration of H^+ in solution is very low. Increasing concentration of H^+ in solution leads to the replacement of M^+ combined in the resin. The cations, M^+_1 that has weaker affinity with resin is replaced by lower concentration of H^+ in solution while the replacement of M^+_2 that has higher affinity with resin needs higher concentration of H^+ in solution.

The coordination ions that need to separate in this experiment are $[Cr(H_2O)_4Cl_2]^+$, $[Cr(H_2O)_5Cl]^{2+}$ and $[Cr(H_2O)_6]^{3+}$, which are present in the solution of $CrCl_3 \cdot 6H_2O$ due to the hydration of Cr^{3+} in a weak acid solution. The corresponding contents of these three coordinate ions depend on time and temperature of the solution storage. In this experiment, $[Cr(H_2O)_4Cl_2]^+$ that has weakest affinity with resin will be replaced firstly by 2×10^{-3} mol·L^{-1} of $HClO_4$ solution, $[Cr(H_2O)_5Cl]^{2+}$ will be replaced by 1.0 mol·L^{-1} of $HClO_4$ solution and $[Cr(H_2O)_6]^{3+}$ that has strongest affinity with resin will be replaced by 3.0 mol·L^{-1} of $HClO_4$ solution at last. The relative content of these ions can be then determined by their UV-Vis spectrum.

III. Apparatus and Reagents

1. Apparatus

Apparatus Name	Specification	Unit and Quantity
UV-Vis spectrometer		1
Beaker	100 mL	1
Glass exchange column	0.5 cm×15 cm	1
Measuring flask 1	100 mL	1
Measuring flask 2	50 mL	1
Burette	10 mL	1

2. Reagents

Reagents Name	Specifications
$CrCl_3 \cdot 6H_2O$	A.R.
HClO$_4$ solution	2×10^{-3} mol·L^{-1}
	1.0 mol·L^{-1}
	3.0 mol·L^{-1}
DI-H$_2$O	Di-ionized water
732 type resin	

IV. Procedures

1. Pretreatment and Column Packing of Resin

(1) Wash commercial resin several times with running water to remove soluble impurity, rinse in DI-H$_2$O for few hours to swell, and wash twice with DI-H$_2$O before rinse in 2 mol·L^{-1} HCl for 10 hours with siring (volume of HCl is 5 times of resin). Then, wash the resin with DI-H$_2$O till pH of washing water is 3.

(2) Load 25 mL of pretreated resin with DI-H$_2$O into exchange column (Note: make sure no space, no air bubbles or drying).

2. Preparation of Solution

(1) Prepare elution solution: dilute perchloric acid (70%) with DI-H$_2$O to make HClO$_4$ solution at concentrations of 0.1 mol·L^{-1}, 1.0 mol·L^{-1} and 3.0 mol·L^{-1}, respectively.

(2) Prepare measured solution: weight out $CrCl_3 \cdot 6H_2O$, dissolve in perchloric acid (70%) and dilute with DI-H$_2$O to make 100 mL of solution containing 0.35 mol·L^{-1} of chromium and 0.002 mol·L^{-1} of HClO$_4$ solution This is the solution of 0.35 mol·L^{-1} $[Cr(H_2O)_4Cl_2]^+$.

3. Preparation of Different Charged Chromium Complex Ions and Determination of Their UV-Vis Spectrum

(1) Add 5 mL of 0.35 mol·L^{-1} $[Cr(H_2O)_4Cl_2]^+$ into ion exchange column and

remove excess solution till it is in the same level of resin. Pour 0.1 mol·L^{-1} HClO$_4$ solution into column to elute [Cr(H$_2$O)$_4$Cl$_2$]$^+$. Elute rate is controlled at 2 drops per second. Collect the solution into a 50 mL of measuring flask once the elute solution shows green. When green color is disappeared in the elution, stop collection and dilute the collected solution in measuring flask with 0.1 mol·L^{-1} of HClO$_4$ solution to make a 50 mL solution. Fill the solution in a cuvette with 1 cm width and determine its spectrum between 350 nm to 700 nm using a VU-Vis spectrometer.

(2) Warm 5 mL of 0.35 mol·L^{-1} [Cr(H$_2$O)$_4$Cl$_2$]$^+$ in a water bath at 55℃ for 2 minutes to transfer [Cr(H$_2$O)$_4$Cl$_2$]$^+$ to [Cr(H$_2$O)$_5$Cl]$^{2+}$. Then add this solution immediately to ion exchange column and remove excess solution till it is in the same level of resin. Add 0.1 mol·L^{-1} of HClO$_4$ into column to elute excess [Cr(H$_2$O)$_4$Cl$_2$]$^+$ (try to figure out whether there is excess [Cr(H$_2$O)$_4$Cl$_2$]$^+$ in the column). After excess [Cr(H$_2$O)$_4$Cl$_2$]$^+$ is eluted completely, add 1.0 mol·L^{-1} of HClO$_4$ into column to elute [Cr(H$_2$O)$_5$Cl]$^{2+}$. Collect the elution and determine its spectrum follow the same procedure as step (1).

(3) Boil 5 mL of 0.35 mol·L^{-1} [Cr(H$_2$O)$_4$Cl$_2$]$^+$ in a oil bath at 110℃ for 5 minutes to transfer [Cr(H$_2$O)$_4$Cl$_2$]$^+$ to [Cr(H$_2$O)$_6$]$^{3+}$. Then cool the solution to room temperature before add to ion exchange column and remove excess solution till it is in the same level of resin. Add 1.0 mol·L^{-1} of HClO$_4$ solution into column to elute excess [Cr(H$_2$O)$_4$Cl$_2$]$^+$ or [Cr(H$_2$O)$_5$Cl]$^{2+}$ (try to figure out whether there is excess [Cr(H$_2$O)$_4$Cl$_2$]$^+$ or [Cr(H$_2$O)$_5$Cl]$^{2+}$ in the column). After excess [Cr(H$_2$O)$_4$Cl$_2$]$^+$ and [Cr(H$_2$O)$_5$Cl]$^{2+}$ is eluted completely, add 3.0 mol·L^{-1} of HClO$_4$ solution into column to elute [Cr(H$_2$O)$_6$]$^{3+}$. Collect the elution and determine its spectrum follow the same procedure as steps (1) and step (2).

4. Separation and Inditification of Chromium Complex Ions in Chromium Chloride

Add 10 mL of CrCl$_6$·6H$_2$O solution prepared several hours before to ion exchange column and remove excess solution till it is in the same level of resin.

Elute $[Cr(H_2O)_4Cl_2]^+$ using 0.1 mol·L^{-1} of HClO$_4$ solution firstly, collect the green elution and determine its spectrum using UV-Vis spectrometer. Then, elute $[Cr(H_2O)_5Cl]^{2+}$ using 1.0 mol·L^{-1} of HClO$_4$ solution, collect the green elution and determine its spectrum using UV-Vis spectrometer. Finally, elute $[Cr(H_2O)_6]^{3+}$ using 3.0 mol·L^{-1} of HClO$_4$ solution, collect the green elution and determine its spectrum using UV-Vis spectrometer.

V. Data recording and Analyzing

(1) Determine the peak wavelength (λ) of absorbance and its molar extinction coefficient (ε) based on the UV-Vis spectrum of coordinate ions.

Coordinate Ion	$[Cr(H_2O)_4Cl_2]^+$	$[Cr(H_2O)_5Cl]^{2+}$	$[Cr(H_2O)_6]^{3+}$
λ			
ε			

(2) Determine the relevant contents of coordinate ions in ion exchange elutions of chromium chloride solution by their UV-Vis spectrum.

Coordinate Ion	$[Cr(H_2O)_4Cl_2]^+$	$[Cr(H_2O)_5Cl]^{2+}$	$[Cr(H_2O)_6]^{3+}$
Relevant content			

VI. Discussion and Exercises

Why do we use perchloric acid instead of hydrochloric acid to elute Cr (III) coordinate ions in ion exchange column?

Part 3　Appendices

Appendix 1　Specification and Selection of Chemical Reagent

The standard of chemical reagent grade varies in different countries. In China, the unified national reagent specification classification refer to the following table:

Specification & Grade	Grade One	Grade Two	Grade Three	Grade Four
Commonly used grade and symbol	Guarantee reagent (G. R.)	Analytical reagent (A. R.)	Chemical pure (C. P.)	Laboratory reagent (L.R.)
Range of application	Accurate analysis and research	General analysis and research	Industrial analysis and chemical experiment	Chemical experiment
Label colors	Green	Red	Blue	Black (Yellow)

Beside the four grades listed in table, there are also other specifications such as organic analytical reagent (O.A.R), micro analytical reagent (M.A.R), standard substance (S.S), spectrum pure, indicator, and industrial reagent and so on.

In general chemical experiment, we usually use chemical pure reagents, and sometimes, laboratory reagents are also used.

Appendix 2 Density and Concentration of Commonly Used Acid or Base Solutions (15℃)

Name of Solution	Density/(g·mL^{-1})	Mass Fraction	Molar Concentration/(mol·L^{-1})
C-H$_2$SO$_4$	1.84	95~96	18
D-H$_2$SO$_4$	1.18	25	3
D-H$_2$SO$_4$	1.06	9	1
C-HCl	1.19	38	12
D-HCl	1.10	20	6
D-HCl	1.03	7	2
C-HNO$_3$	1.40	65	14
D-HNO$_3$	1.20	32	6
D-HNO$_3$	1.07	12	2
C-H$_3$PO$_4$	1.7	85	15
D-H$_3$PO$_4$	1.05	9	1
D-HClO$_4$	1.12	19	2
C-HF	1.13	40	23
HBr	1.38	40	7
HI	1.70	57	7.5
Iced HAc	1.05	99~100	17.5
D-HAc	1.04	35	6
D-HAc	1.02	12	2
C-NaOH	1.36	33	11
D-NaOH	1.09	8	2
C-NH$_3$ (aq)	0.88	35	18
C-NH$_3$ (aq)	0.91	25	13.5
D-NH$_3$ (aq)	0.96	11	6
D-NH$_3$ (aq)	0.99	3.5	2

* C-: Concentrated; D-: Diluted.

Appendix 3   Color of Common Ions

1. Colorless cations

Ag^+, Cd^{2+}, K^+, Ca^{2+}, As^{3+}, Pb^{2+}, Zn^{2+}, Na^+, Sr^{2+}, As^{5+}, Hg^{2+}, Bi^{3+}, NH_4^+, Ba^{2+}, Sb^{5+}, Sb^{3+}, Mg^{2+}, Al^{3+}, Sn^{2+}, Sn^{4+}.

2. Colored cations

Fe^{3+}: yellow or red brown; Fe^{2+}: light green; Cr^{3+}: green or purple; Co^{2+}: rose; Ni^{2+}: green; Cu^{2+}: light blue.

3. Colorless anions

SO_4^{2-}, PO_4^{3-}, F^-, SCN^-, $C_2O_4^{2-}$, MoO_4^{2-}, SO_3^{2-}, Cl^-, NO_3^-, S^{2-}, $S_2O_3^{2-}$, Br^-, NO_2^-, ClO_3^{2-}, CO_3^{2-}, SiO_3^{2-}, HCO_3^{2-}, PbI_4^{2-}.

4. Colored anions

$Cr_2O_7^{2-}$: orange; CrO_4^{2-}: yellow; CrO_2^-: green; MnO_4^-: purple red; MnO_4^{2-}: green; $[Fe(CN)_6]^{3-}$: red brown; $[Fe(CN)_6]^{4-}$: yellow-green; $[CuCl_4]^{2-}$: yellow.

Appendix 4 International Relative Atomic Mass

Atomic Number	Symbol	Element Name	Atomic Mass	Atomic Number	Symbol	Element Name	Atomic Mass
1	H	Hydrogen	1.008	28	Ni	Nickel	58.69
2	He	Helium	4.003	29	Cu	Copper	63.55
3	Li	Lithium	6.941	30	Zn	Zinc	65.39
4	Be	Beryllium	9.012	31	Ga	Gallium	69.72
5	B	Boron	10.81	32	Ge	Germanium	72.61
6	C	Carbon	12.01	33	As	Arsenic	74.92
7	N	Nitrogen	14.007	34	Se	Selenium	78.96
8	O	Oxygen	15.999	35	Br	Bromine	79.90
9	F	Fluorine	18.998	36	Kr	Krypton	83.80
10	Ne	Neon	20.18	37	Rb	Rubidium	85.47
11	Na	Sodium	22.99	38	Sr	Strontium	87.62
12	Mg	Magnesium	24.305	39	Y	Yttrium	88.91
13	Al	Aluminum	26.98	40	Zr	Zirconium	91.22
14	Si	Silicon	28.09	41	Nb	Niobium	92.91
15	P	Phosphorus	30.97	42	Mo	Molybdenum	95.94
16	S	Sulfur	32.07	43	^{99}Tc	Technetium	98.9
17	Cl	Chlorine	35.45	44	Ru	Ruthenium	101.1
18	Ar	Argon	39.95	45	Rh	Rhodium	102.9
19	K	Potassium	39.10	46	Pd	Palladium	106.4
20	Ca	Calcium	40.08	47	Ag	Silver	107.9
21	Sc	Scandium	44.96	48	Cd	Cadmium	112.4
22	Ti	Titanium	47.87	49	In	Indium	114.8
23	V	Vanadium	50.94	50	Sn	Tin	118.7
24	Cr	Chromium	52.00	51	Sb	Antimony	121.8
25	Mn	Manganese	54.94	52	Te	Tellurium	127.6
26	Fe	Iron	55.845	53	I	Iodine	126.9
27	Co	Cobalt	58.93	54	Xe	Xenon	131.3

Atomic Number	Symbol	Element Name	Atomic Mass	Atomic Number	Symbol	Element Name	Atomic Mass
55	Cs	Cerium	132.9	68	Er	Erbium	167.3
56	Ba	Barium	137.3	69	Tm	Thulium	168.9
57	La	Lanthanum	138.9	70	Yb	Ytterbium	173.0
58	Ce	Cerium	140.1	71	Lu	Lutetium	175.0
59	Pr	Praseodymium	140.9	72	Hf	Hafnium	178.5
60	Nd	Niobium	144.2	73	Ta	Tantalum	180.9
61	^{145}Pm	Promethium	144.9	74	W	Tungsten	183.8
62	Sm	Samarium	150.4	75	Re	Rhenium	186.2
63	Eu	Europium	152.0	76	Os	Osmium	190.2
64	Gd	Gadolinium	157.3	77	Ir	Iridium	192.2
65	Tb	Terbium	158.9	78	Pt	Platinum	195.1
66	Dy	Dysprosium	162.5	79	Au	Gold	197.0
67	Ho	Holmium	164.9	80	Hg	Mercury	200.6

Continued

Appendix 5 Standard Electrode Potential

In acid solutions (298K)

Electrode Pair	Electrode Reaction Equation	$\varphi^\ominus$ /V
Li^+/Li	$Li^+ + e^- = Li$	−3.0401
Cs^+/Cs	$Cs^+ + e^- = Cs$	−3.026
Rb^+/Rb	$Rb^+ + e^- = Rb$	−2.98
K^+/K	$K^+ + e^- = K$	−2.931
Ba^{2+}/Ba	$Ba^{2+} + 2e^- = Ba$	−2.912
Sr^{2+}/Sr	$Sr^{2+} + 2e^- = Sr$	−2.89
Ca^{2+}/Ca	$Ca^{2+} + 2e^- = Ca$	−2.868
Na^+/Na	$Na^+ + e^- = Na$	−2.71
La^{3+}/La	$La^{3+} + 3e^- = La$	−2.379
Mg^{2+}/Mg	$Mg^{2+} + 2e^- = Mg$	−2.372
Ce^{3+}/Ce	$Ce^{3+} + 3e^- = Ce$	−2.336
H_2/H^-	$H_2(g) + 2e^- = 2H^-$	−2.23
AlF_6^{3-}/F^-	$AlF_6^{3-} + 3e^- = Al + 6F^-$	−2.069
Th^{4+}/Th	$Th^{4+} + 4e^- = Th$	−1.899
Be^{2+}/Be	$Be^{2+} + 2e^- = Be$	−1.847
U^{3+}/U	$U^{3+} + 3e^- = U$	−1.798
Al^{3+}/Al	$Al^{3+} + 3e^- = Al$	−1.662
Ti^{2+}/Ti	$Ti^{2+} + 2e^- = Ti$	−1.630
$ZrO_2/Zr/$	$ZrO_2 + 4H^+ + 4e^- = Zr + 2H_2O$	−1.553
$[SiF_6]^{2-}/Si^-$	$[SiF_6]^{2-} + 4e^- = Si + 6F^-$	−1.24
Mn^{2+}/Mn	$Mn^{2+} + 2e^- = Mn$	−1.185
Cr^{2+}/Cr	$Cr^{2+} + 2e^- = Cr$	−0.913
Ti^{3+}/Ti^{2+}	$Ti^{3+} + e^- = Ti^{2+}$	−0.9
H_3BO_3/B	$H_3BO_3 + 3H^+ + 3e^- = B + 3H_2O$	−0.8698
TiO_2/Ti	$TiO_2 + 4H^+ + 4e^- = Ti + 2H_2O$	−0.86
Te/H_2Te	$Te + 2H^+ + 2e^- = H_2Te$	−0.793
Zn^{2+}/Zn	$Zn^{2+} + 2e^- = Zn$	−0.7618
Ta_2O_5/Ta	$Ta_2O_5 + 10H^+ + 10e^- = 2Ta + 5H_2O$	−0.750

Continued

Electrode Pair	Electrode Reaction Equation	$\varphi^{\ominus}$ /V
Cr^{3+}/Cr	$Cr^{3+}+3e^- \rightleftharpoons Cr$	−0.744
Nb_2O_5/Nb	$Nb_2O_5+10H^++10e^- \rightleftharpoons 2Nb+5H_2O$	−0.644
As/AsH_3	$As+3H^++3e^- \rightleftharpoons AsH_3$	−0.608
U^{4+}/U^{3+}	$U^{4+}+e^- \rightleftharpoons U^{3+}$	−0.607
Ga^{3+}/Ga	$Ga^{3+}+3e^- \rightleftharpoons Ga$	−0.549
$CO_2/H_2C_2O_4$	$2CO_2+2H^++2e^- \rightleftharpoons H_2C_2O_4$	−0.49
Fe^{2+}/Fe	$Fe^{2+}+2e^- \rightleftharpoons Fe$	−0.447
Cr^{3+}/Cr^{2+}	$Cr^{3+}+e^- \rightleftharpoons Cr^{2+}$	−0.407
Cd^{2+}/Cd	$Cd^{2+}+2e^- \rightleftharpoons Cd$	−0.4030
Se/H_2Se	$Se+2H^++2e^- \rightleftharpoons H_2Se\ (aq)$	−0.399
PbI_2/Pb	$PbI_2+2e^- \rightleftharpoons Pb+2I^-$	−0.365
Eu^{3+}/Eu^{2+}	$Eu^{3+}+e^- \rightleftharpoons Eu^{2+}$	−0.36
Pb^{2+}/Pb	$PbSO_4+2e^- \rightleftharpoons Pb+SO_4^{2-}$	−0.3588
In^{3+}/In	$In^{3+}+3e^- \rightleftharpoons In$	−0.3382
Tl^+/Tl	$Tl^++e^- \rightleftharpoons Tl$	−0.336
Co^{2+}/Co	$Co^{2+}+2e^- \rightleftharpoons Co$	−0.28
H_3PO_4/H_3PO_3	$H_3PO_4+2H^++2e^- \rightleftharpoons H_3PO_3+H_2O$	−0.276
Pb^{2+}/Pb	$PbCl_2+2e^- \rightleftharpoons Pb+2Cl^-$	−0.2675
Ni^{2+}/Ni	$Ni^{2+}+2e^- \rightleftharpoons Ni$	−0.257
V^{3+}/V^{2+}	$V^{3+}+e^- \rightleftharpoons V^{2+}$	−0.255
H_2GeO_3/Ge	$H_2GeO_3+4H^++4e^- \rightleftharpoons Ge+3H_2O$	−0.182
AgI/Ag	$AgI+e^- \rightleftharpoons Ag+I^-$	−0.15224
Sn^{2+}/Sn	$Sn^{2+}+2e^- \rightleftharpoons Sn$	−0.1375
Pb^{2+}/Pb	$Pb^{2+}+2e^- \rightleftharpoons Pb$	−0.1262
CO_2/CO	$CO_2\ (g)+2H^++2e^- \rightleftharpoons CO+H_2O$	−0.12
Hg_2I_2/Hg	$Hg_2I_2+2e^- \rightleftharpoons 2Hg+2I^-$	−0.0405
Fe^{3+}/Fe	$Fe^{3+}+3e^- \rightleftharpoons Fe$	−0.037
$2H^+/H_2$	$2H^++2e^- \rightleftharpoons H_2$	+0.0000
$AgBr/Ag$	$AgBr+e^- \rightleftharpoons Ag+Br^-$	+0.07133
$S_4O_6^{2-}/S_2O_3^{2-}$	$S_4O_6^{2-}+2e^- \rightleftharpoons 2S_2O_3^{2-}$	+0.08
TiO^{2+}/Ti^{3+}	$TiO^{2+}+2H^++e^- \rightleftharpoons Ti^{3+}+H_2O$	+0.1
$S/H_2S\ (aq)$	$S+2H^++2e^- \rightleftharpoons H_2S\ (aq)$	+0.142
Sn^{4+}/Sn^{2+}	$Sn^{4+}+2e^- \rightleftharpoons Sn^{2+}$	+0.151
Sb_2O_3/Sb	$Sb_2O_3+6H^++6e^- \rightleftharpoons 2Sb+3H_2O$	+0.152

Continued

Electrode Pair	Electrode Reaction Equation	$\varphi^{\ominus}$ /V
Cu^{2+}/Cu^{+}	$Cu^{2+}+e^{-} =\!=\!= Cu^{+}$	+0.153
SO_4^{2-}/H_2SO_3	$SO_4^{2-}+4H^{+}+2e^{-} =\!=\!= H_2SO_3+H_2O$	+0.172
SbO^{+}/Sb	$SbO^{+}+2H^{+}+3e^{-} =\!=\!= Sb+H_2O$	+0.212
$AgCl/Ag$	$AgCl+e^{-} =\!=\!= Ag+Cl^{-}$	+0.22233
$HAsO_2/As$	$HAsO_2+3H^{+}+3e^{-} =\!=\!= As+2H_2O$	+0.248
Hg_2Cl_2/Hg	$Hg_2Cl_2+2e^{-} =\!=\!= 2Hg+2Cl^{-}$ (饱和 KCl)	+0.26808
Cu^{2+}/Cu	$Cu^{2+}+2e^{-} =\!=\!= Cu$	+0.3419
ReO_4^{-}/Re	$ReO_4^{-}+8H^{+}+7e^{-} =\!=\!= Re+4H_2O$	+0.368
Ag_2CrO_4/Ag	$Ag_2CrO_4+2e^{-} =\!=\!= 2Ag+CrO_4^{2-}$	+0.4470
H_2SO_3/S	$H_2SO_3+4H^{+}+4e^{-} =\!=\!= S+3H_2O$	+0.449
Cu^{+}/Cu	$Cu^{+}+e^{-} =\!=\!= Cu$	+0.521
I_2/I^{-}	$I_2+2e^{-} =\!=\!= 2I^{-}$	+0.5355
I_3^{-}/I^{-}	$I_3^{-}+2e^{-} =\!=\!= 3I^{-}$	+0.536
O_2/H_2O_2	$O_2+2H^{+}+2e^{-} =\!=\!= H_2O_2$	+0.695
Fe^{3+}/Fe^{2+}	$Fe^{3+}+e^{-} =\!=\!= Fe^{2+}$	+0.771
Hg_2^{2+}/Hg	$Hg_2^{2+}+2e^{-} =\!=\!= 2Hg$	+0.7973
Ag^{+}/Ag	$Ag^{+}+e^{-} =\!=\!= Ag$	+0.7996
NO_3^{-}/N_2O_4	$2NO_3^{-}+4H^{+}+2e^{-} =\!=\!= N_2O_4+2H_2O$	+0.803
Hg^{2+}/Hg	$Hg^{2+}+2e^{-} =\!=\!= Hg$	+0.851
(quartz) SiO_2/Si	(quartz) $SiO_2+4H^{+}+4e^{-} =\!=\!= Si+2H_2O$	+0.857
Hg^{2+}/Hg_2^{2+}	$2Hg^{2+}+2e^{-} =\!=\!= Hg_2^{2+}$	+0.920
$Pd^{2+}+2e^{-} =\!=\!= Pd$	$Pd^{2+}+2e^{-} =\!=\!= Pd$	+0.951
NO_3^{-}/NO	$NO_3^{-}+4H^{+}+3e^{-} =\!=\!= NO+2H_2O$	+0.957
HNO_2/NO	$HNO_2+H^{+}+e^{-} =\!=\!= NO+H_2O$	+0.983
HIO/I^{-}	$HIO+H^{+}+2e^{-} =\!=\!= I^{-}+H_2O$	+0.987
$[AuCl_4]^{-}/Au$	$[AuCl_4]^{-}+3e^{-} =\!=\!= Au+4Cl^{-}$	+1.002
H_6TeO_6/TeO_2	$H_6TeO_6+2H^{+}+2e^{-} =\!=\!= TeO_2+4H_2O$	+1.02
IO_3^{-}/I^{-}	$IO_3^{-}+6H^{+}+6e^{-} =\!=\!= I^{-}+3H_2O$	+1.085
$Br_2(aq)/Br^{-}$	$Br_2(aq)+2e^{-} =\!=\!= 2Br^{-}$	+1.0873
Pt^{2+}/Pt	$Pt^{2+}+2e^{-} =\!=\!= Pt$	+1.18
ClO_4^{-}/ClO_3^{-}	$ClO_4^{-}+2H^{+}+2e^{-} =\!=\!= ClO_3^{-}+H_2O$	+1.189
IO_3^{-}/I_2	$2IO_3^{-}+12H^{+}+10e^{-} =\!=\!= I_2+6H_2O$	+1.195
$ClO_3^{-}/HClO_2$	$ClO_3^{-}+3H^{+}+2e^{-} =\!=\!= HClO_2+H_2O$	+1.214
MnO_2/Mn^{2+}	$MnO_2+4H^{+}+2e^{-} =\!=\!= Mn^{2+}+2H_2O$	+1.224

Continued

Electrode Pair	Electrode Reaction Equation	$\varphi^{\ominus}$ /V
O_2/H_2O	$O_2+4H^++4e^- \rightleftharpoons 2H_2O$	+1.229
Tl^{3+}/Tl^+	$Tl^{3+}+2e^- \rightleftharpoons Tl^+$	+1.252
$Cr_2O_7^{2-}/Cr^{3+}$	$Cr_2O_7^{2-}+14H^++6e^- \rightleftharpoons 2Cr^{3+}+7H_2O$	+1.33
$HCrO_4^-/Cr^{3+}$	$HCrO_4^-+7H^++3e^- \rightleftharpoons Cr^{3+}+4H_2O$	+1.350
$Cl_2(g)/Cl^-$	$Cl_2(g)+2e^- \rightleftharpoons 2Cl^-$	+1.35827
ClO_4^-/Cl^-	$ClO_4^-+8H^++8e^- \rightleftharpoons Cl^-+4H_2O$	+1.389
Au^{3+}/Au^+	$Au^{3+}+2e^- \rightleftharpoons Au^+$	+1.401
BrO_3^-/Br^-	$BrO_3^-+6H^++6e^- \rightleftharpoons Br^-+3H_2O$	+1.423
HIO/I_2	$2HIO+2H^++2e^- \rightleftharpoons I_2+2H_2O$	+1.439
ClO_3^-/Cl^-	$ClO_3^-+6H^++6e^- \rightleftharpoons Cl^-+3H_2O$	+1.451
PbO_2/Pb^{2+}	$PbO_2+4H^++2e^- \rightleftharpoons Pb^{2+}+2H_2O$	+1.455
Au^{3+}/Au	$Au^{3+}+3e^- \rightleftharpoons Au$	+1.498
MnO_4^-/Mn^{2+}	$MnO_4^-+8H^++5e^- \rightleftharpoons Mn^{2+}+4H_2O$	+1.507
Mn^{3+}/Mn^{2+}	$Mn^{3+}+e^- \rightleftharpoons Mn^{2+}$	+1.5415
NO/N_2O	$2NO+2H^++2e^- \rightleftharpoons N_2O+H_2O$	+1.591
NiO_2/Ni^{2+}	$NiO_2+4H^++2e^- \rightleftharpoons Ni^{2+}+2H_2O$	+1.678
MnO_4^-/MnO_2	$MnO_4^-+4H^++3e^- \rightleftharpoons MnO_2+2H_2O$	+1.679
Au^+/Au	$Au^++e^- \rightleftharpoons Au$	+1.692
Ce^{4+}/Ce^{3+}	$Ce^{4+}+e^- \rightleftharpoons Ce^{3+}$	+1.72
H_2O_2/H_2O	$H_2O_2+2H^++2e^- \rightleftharpoons 2H_2O$	+1.776
Co^{3+}/Co^{2+} (2mol·L^{-1} H$_2$SO$_4$)	$Co^{3+}+e^- \rightleftharpoons Co^{2+}$ (2mol·L^{-1} H$_2$SO$_4$)	+1.83
Ag^{2+}/Ag^+	$Ag^{2+}+e^- \rightleftharpoons Ag^+$	+1.980
$S_2O_8^{2-}/SO_4^{2-}$	$S_2O_8^{2-}+2e^- \rightleftharpoons 2SO_4^{2-}$	+2.010
O_3/O_2	$O_3+2H^++2e^- \rightleftharpoons O_2+H_2O$	+2.076
$O(g)/H_2O$	$O(g)+2H^++2e^- \rightleftharpoons H_2O$	+2.421
F_2/F^-	$F_2+2e^- \rightleftharpoons 2F^-$	+2.866

Appendix 6　Saturated Vapor Pressure of Water at Different Temperature

T/°C	P/kPa	T/°C	P/kPa	T/°C	P/kPa
0	0.6125	34	5.320	68	28.56
1	0.6568	35	5.623	69	29.83
2	0.7058	36	5.942	70	31.16
3	0.7580	37	6.275	71	32.52
4	0.8134	38	6.625	72	33.95
5	0.8724	39	6.992	73	35.43
6	0.9350	40	7.376	74	35.96
7	1.002	41	7.778	75	38.55
8	1.073	42	8.200	76	40.19
9	1.148	43	8.640	77	41.88
10	1.228	44	9.101	78	43.64
11	1.312	45	9.584	79	45.47
12	1.402	46	10.09	80	47.35
13	1.497	47	10.61	81	49.29
14	1.598	48	11.16	82	51.32
15	1.705	49	11.74	83	53.41
16	1.818	50	12.33	84	55.57
17	1.937	51	12.96	85	57.81
18	2.064	52	13.61	86	60.12
19	2.197	53	14.29	87	62.49
20	2.338	54	15.00	88	64.94
21	2.487	55	15.74	89	67.48
22	2.644	56	16.51	90	70.10
23	2.809	57	17.31	91	72.80
24	2.985	58	18.14	92	75.60

Continued

T/°C	P/kPa	T/°C	P/kPa	T/°C	P/kPa
25	3.167	59	19.01	93	78.48
26	3.361	60	19.92	94	81.45
27	3.565	61	20.86	95	84.52
28	3.780	62	21.84	96	87.67
29	4.006	63	22.85	97	90.94
30	4.248	64	23.91	98	94.30
31	4.493	65	25.00	99	97.76
32	4.755	66	26.14	100	101.30
33	5.030	67	27.33		

Appendix 7 Stability Constant of Some Coordinated Ions

Coordinated Ions	$K_s/[K_s]$	$Lg(K_s/[K_s])$
$[Ag(CN)_2]^-$	1.26×10^{21}	21.2
$[Ag(NH_3)_2]^+$	1.12×10^7	7.05
$[Ag(S_2O_3)_2]^{3-}$	2.89×10^{13}	13.46
$[AgCl_2]^-$	1.10×10^5	5.04
$[AgBr_2]^-$	2.14×10^7	7.33
$[AgI_2]^-$	5.50×10^{11}	11.74
$[Ag(py)_2]^+$	1.0×10^{10}	10.0
$[Co(NH_3)_6]^{2+}$	1.29×10^5	5.11
$[Cu(CN)_2]^-$	1.0×10^{24}	24
$[Cu(SCN)_2]^-$	1.52×10^5	5.18
$[Cu(NH_3)_2]^+$	7.24×10^{10}	10.86
$[Cu(NH_3)_4]^{2+}$	2.09×10^{13}	13.32
$[Cu(P_2O_7)_2]^{6-}$	1.0×10^9	9.0
$[FeF_6]^{3-}$	2.04×10^{14}	14.31
$[Fe(CN)_6]^{3-}$	1.0×10^{42}	42
$[Hg(CN)_4]^{2-}$	2.51×10^{41}	41.4
$[HgI_4]^{2-}$	6.76×10^{29}	29.83
$[HgBr_4]^{2-}$	1.0×10^{21}	21.00
$[HgCl_4]^{2-}$	1.17×10^{15}	15.07
$[Ni(NH_3)_6]^{2+}$	5.5×10^8	8.74
$[Ni(en)_3]^{2+}$	2.14×10^{18}	18.33
$[Zn(CN)_4]^{2-}$	5.0×10^{16}	16.7
$[Zn(NH_3)_4]^{2+}$	2.87×10^9	9.46
$[Zn(en)_2]^{2+}$	6.76×10^{10}	10.83

Appendix 8 Boiling Point of Water at Different Pressure

p/kPa	T_b/°C
0.611	0.01 (Triple point)
50.66	80.9
101.3	100.0
202.7	119.6
304.0	132.9
405.3	142.9
506.6	151.1
1013.3	179.0
1519.9	197.4
2026.5	211.4
2533.1	222.9
22120.0	374.1 (Critical temperature)

Appendix 9　Density of Water at Different Temperature

$T/°C$	$\rho/(g \cdot mL^{-1})$
−10	0.998 12
−5	0.999 27
0	0.999 64
4	0.999 97
5	0.999 96
10	0.999 70
18	0.998 59
20	0.998 20
25	0.997 04
30	0.995 64
40	0.992 21
50	0.988 04
60	0.983 21
70	0.977 78
80	0.971 80
85	0.968 62
90	0.965 34
95	0.961 89
100	0.958 35
110	0.950 97

* These data comes from *Lange's Handbook of Chemistry*, 11th Ed., and calculated according to 1 atm= 101.325 kPa.

Appendix 10　Relationship Between pH and Temperature of Buffer Solutions

$T/°C$	Potassium Acid Phthalate (0.005 mol·kg^{-1})	Phosphate Mixture (0.025 mol·kg^{-1})	Sodium Borate (0.01 mol·kg^{-1})
5	4.00	6.95	9.39
10	4.00	6.92	9.33
15	4.00	6.90	9.28
20	4.00	6.88	9.23
25	4.00	6.86	9.18
30	4.01	6.85	9.14
35	4.02	6.84	9.11
40	4.03	6.84	9.07
45	4.04	6.84	9.04
50	4.06	6.83	9.03

Appendix 11 Dissociation Constant of Some Common Weak Electrolytes (298.15 K)

Electrolyte	Chemical Formula	Dissociation Equilibrium	Dissociation Constant
Acetic acid	CH_3COOH	$CH_3COOH \rightleftharpoons H^+ + CH_3COO^-$	$K_a = 1.74 \times 10^{-5}$
Carbonic acid	H_2CO_3	$H_2CO_3 \rightleftharpoons H^+ + HCO_3^-$ $HCO_3^- \rightleftharpoons H^+ + CO_3^{2-}$	$K_{a1} = 4.47 \times 10^{-7}$ $K_{a2} = 4.68 \times 10^{-11}$
Hydrogen sulfuric acid	H_2S	$H_2S \rightleftharpoons H^+ + HS^-$ $HS^- \rightleftharpoons H^+ + S^{2-}$	$K_{a1} = 8.91 \times 10^{-8}$ $K_{a1} = 1.0 \times 10^{-19}$
Oxalic acid	$H_2C_2O_4$	$H_2C_2O_4 \rightleftharpoons H^+ + HC_2O_4^-$ $HC_2O_4^- \rightleftharpoons H^+ + C_2O_4^{2-}$	$K_{a1} = 5.89 \times 10^{-2}$ $K_{a2} = 6.46 \times 10^{-5}$
Phosphoric acid	H_3PO_4	$H_3PO_4 \rightleftharpoons H^+ + H_2PO_4^-$ $H_2PO_4^- \rightleftharpoons H^+ + HPO_4^{2-}$ $HPO_4^{2-} \rightleftharpoons H^+ + PO_4^{3-}$	$K_{a1} = 6.92 \times 10^{-3}$ $K_{a2} = 6.17 \times 10^{-8}$ $K_{a31} = 4.79 \times 10^{-13}$
Ammonia	NH_3	$NH_3 + H_2O \rightleftharpoons NH_4^+ + OH^-$	$K_b = 1.78 \times 10^{-5}$
Aniline	$C_6H_5NH_2$	$C_6H_5NH_2 + H_2O \rightleftharpoons C_6H_5NH_3^+ + OH^-$	$K_b = 4.2 \times 10^{-10}$

Appendix 12 Solubility Product of Some Common Substances

Insoluble Matter	Solubility Product
AgCl	1.77×10^{-10}
AgBr	5.35×10^{-13}
AgI	8.51×10^{-17}
Ag_2CrO_4	1.12×10^{-12}
Ag_2S	6.69×10^{-50} (α)
	1.09×10^{-49} (β)
CuS	1.27×10^{-36}
$Fe(OH)_3$	2.64×10^{-39}
$Fe(OH)_2$	4.87×10^{-17}
$Mg(OH)_2$	5.61×10^{-12}
$Mn(OH)_2$	2.06×10^{-13}
MnS	4.65×10^{-14}
ZnS	2.93×10^{-25}
CdS	1.40×10^{-29}

Appendix 13 Symbols, Values and SI Units of Some Constants

Physical Quality	Symbol	Value	Unit
Elementary charge	e	$1.602\,2 \times 10^{-19}$	C
Rest mass of electron	m_e	$9.109\,38 \times 10^{-31}$	kg
Rest mass of proton	m_p	$1.672\,62 \times 10^{-27}$	kg
Light speed in vacuum	c, c_0	$2.997\,9 \times 10^8$	m·s^{-1}
Magnetic permeability in vacuum	μ_0	$12.566\,37 \times 10^{-7}\ (4\pi \times 10^{-7})$	N·A^{-2}
Capacitance ratio in vacuum	ε_0	$8.854\,187\,817 \times 10^{-12}$	F·m^{-1}
Planck constant	h	$6.626\,069 \times 10^{-34}$	J·s
Rydberg constant	R	$1.097\,4 \times 10^7$	m^{-1}
Avogadro constant	N_A	$6.022\,14 \times 10^{23}$	mol^{-1}
Faraday constant	F	$9.648\,534 \times 10^4$	C·mol^{-1}
Molar gas constant	R	$8.314\,47$	J·mol^{-1}·K^{-1}
Boltzmann constant	k	$1.380\,65 \times 10^{-23}$	J·K^{-1}

Class_____ Group No._____ Name_____ Partner Name _____ Date_____

Part 4   Experimental Reports

Experimental Report 1

I. Objects

II. Principles

III. Data Recording

Mass of Al: $W_{Al} =$ _____ g.

Volume of eudiometer before reaction: $V_1 =$ _____ mL = _____ m³.

Volume of eudiometer after reaction: $V_2 =$ _____ mL = _____ m³.

Room temperature: $t =$ _____ ℃ = _____ K.

Atmosphere pressure: $p_{atm} =$ _____ kPa = _____ Pa.

The partial pressure of water vapor: $p_{water} =$ _____ Pa.

IV. Data Analyzing

Volume of H$_2$: $V(H_2) = V_2 - V_1 =$ _____ m³.

The partial pressure of H$_2$: $p(H_2) = p_{atm} - p(H_2O) =$ _____ Pa.

The H$_2$ amonut of substance: $n(H_2) = 3W_{Al}/2M_{Al} =$ _____ mol.

Ideal gas constant: $R = p(H_2)V(H_2)/n(H_2)T$

= _____

= _____ J·mol⁻¹·K⁻¹.

Ideal gas constant in theory: $R_{theory} = 8.314$ J·mol⁻¹·K⁻¹(Pa·m³·mol⁻¹·K⁻¹)

= 0.082 atm·L·mol⁻¹·K⁻¹

= 62400 mmHg·mL·mol⁻¹·K⁻¹.

V. Relative Error

Relative Error: $= |R_{theory} - R_{real}|/R_{theory} \times 100\%$

= _____

= _____ .

Notice: The relative error of this experiment is required less than 10%. If it is higher than 10%, you should reform the experiment at once.

VI. Discussion and Exercises

Class_____ Group No._____ Name_____ Partner Name _____ Date_____

Experimental Report 2

I. Objects

II. Principles

III. Data Recording and Analyzing

1. Data Recording

$CuSO_4$ concentration = _____ mol·L^{-1}, amount = _____ mL, Zn powder = _____ g.

The relation between reaction time and temperature

Time/min							
Temperature/℃							

Time/min							
Temperature/℃							

Time/min							
Temperature/℃							

2. Data Analyzing

(1) Determine ΔT by extrapolation:

(2) Data analyzing:

$\Delta T =$ _____.

$n =$ _____.

$\Delta_r H_m^\ominus = \Delta T C d V / 1000 n =$ _____.

IV. Error Analyzing

V. Discussion and Exercises

Class_____ Group No._____ Name_____ Partner Name_____ Date_____

Experimental Report 3

I. Objects

II. Principles

III. Rate Affect Factors: Concentration and Catalyst

	Experiment Number	1	2	3	4
Amount/mL	KI solution (0.01 mol·L^{-1})	10	5	10	5
	KBrO$_3$ solution (0.04 mol·L^{-1})	10	10	5	10
	Na$_2$S$_2$O$_3$ solution (0.001 mol·L^{-1})	10	10	10	10
	HCl solution (0.1 mol·L^{-1})	10	10	10	10
	Starch solution (0.2%)	2	2	2	2
	KNO$_3$ solution (0.01 mol·L^{-1})	0	5	0	5
	KNO$_3$ solution (0.04 mol·L^{-1})	0	0	5	0
	(NH$_4$)$_6$Mo$_7$O$_{24}$·4H$_2$O solution (0.06 mol·L^{-1})	0	0	0	1 drop
Initial concentration /mol·L^{-1}	KI solution				
	KBrO$_3$ solution				
	Na$_2$S$_2$O$_3$ solution				
Reaction Time (t/s)					
Reaction Rate k					

Discuss following questions based on experimental data:

(1) How does concentration affect reaction rate and rate constant?

(2) How does catalyst affect reaction rate and rate constant?

IV. Temperature Affect Reaction Rate

Items	Reagents				
	KI	KBrO$_3$	Na$_2$S$_2$O$_3$	HCl	Starch
Concentration before mix/ mol·L^{-1}	0.01	0.04	0.001	0.1	0.2%
Volume before mix /mL	5	5	5	5	2
Concentration after mix/ mol·L^{-1}					

	Actual Reaction Temperature	Room Temperature (r.t.)	r.t.+ 10℃	r.t.+ 20℃	r.t.+ 30℃
Data recording		K	K	K	K
	Reaction Time/s				
	$k = \dfrac{\Delta c(S_2O_3^{2-})}{6\Delta t c(BrO_3^-)c(I^-)}$				
	lg$k$				
	1/T				

Plotting a slope which shows lg$k$ as a function of 1/T (1/T as x axis, lg$k$ as y axis).

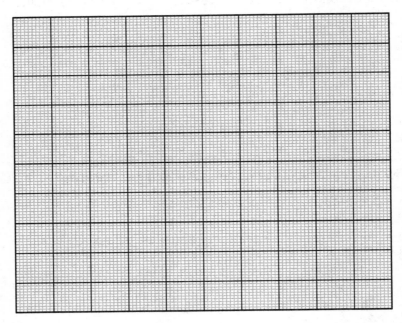

V. Calculation of activation energy (kJ·mol⁻¹)

(1) Value of a, b.

$a = ($ $) - ($ $) = $ _____ .

$b = ($ $) - ($ $) = $ _____ .

(2) Activation energy E.

$E = -2.303 \times R \times (a/b) = $ _____ kJ·mol⁻¹.

VI. Discussion and Exercises

Class_____Group No._____Name_____Partner_____Date_____

Experimental Report 4

I. Objects

II. Principles

III. Data Recording

Items	Phenomenon	Reaction Equations and Illuminations
Galvanic cell	When the salt bridge was used: Potential: When the salt bridge was not used:	Positive electrode: Negative electrode: Functions of salt bridge: The reason why the experimental potential was different with the theoretical value
Electroplating		Anode: Cathode:
Electrochemical corrosion of metal (Zinc particle)	Before the copper rod was inserted:	
	After the copper rod was inserted:	Anode: Cathode:
Electrochemical corrosion of metal (Zinc-coated iron)		Anode: Cathode: Reaction equation for the color change:
Influence of the corrosion inhibitor and cathode protection method (Effect of the corrosion inhibitor)	Before the urotropine was added:	Corrosion reaction equation:
	After the urotropine was added:	Reaction equation for the color change: Conclusion:
Influence of the corrosion inhibitor and cathode protection method (Cathode protection)	Anode: Cathode:	Anode: Cathode: Conclusion:

IV. Discussion and Exercises

Class_____ Group No._____ Name_____ Partner_____ Date_____

Experimental Report 5

I. Objects

II. Principles

III. Data Recording

Items	Experiment Steps and Phenomenons	Reaction Equations and Conlusions
Formation and dissociation of Ag (I) coordinated ion		
Formation and dissociation of Fe (III) coordinated ion		
Formation and color change of Ni (II) coordinated compound		
Properties of silver halide	<table><tr><td></td><td>HNO$_3$</td><td>NH$_3$·H$_2$O</td><td>Na$_2$S$_2$O$_3$</td></tr><tr><td>AgCl</td><td></td><td></td><td></td></tr><tr><td>AgBr</td><td></td><td></td><td></td></tr><tr><td>AgI</td><td></td><td></td><td></td></tr></table>	
Oxidation property of KMnO$_4$		
Reciprocal transformation between K$_2$Cr$_2$O$_7$ and K$_2$CrO$_4$		
The redox properties of H$_2$O$_2$		

IV. Discussion and Exercises

Class_____ Group No._____ Name_____ Partner_____ Data_____

Experimental Report 6

I. Objects

II. Principles

III. Data recording

Room temperature: _____ °C   Wavelength λ: _____ nm   Date: _____

No.	Bromothymol Blue Solution/mL	NaH$_2$PO$_4$ Solution /mL	K$_2$HPO$_4$ Solution /mL	Other Reagents	pH	A	$\dfrac{[\text{In}^-]}{[\text{HIn}]} = \dfrac{A^0_{\text{HIn}} - A_x}{A_x - A^0_{\text{In}^-}}$	p$K^\ominus_a$
1	1.00	0	0	4 drops of HCl solution		0		
2	1.00	2.5	0.5					
3	1.00	5.0	2.5					
4	1.00	2.5	5.0					
5	1.00	0.5	2.5					
6	1.00	0.5	5.0					
7	1.00	0	0	5 drops of NaOH solution				

Making the curve of pH and $\lg \dfrac{A^0_{\text{HIn}} - A_x}{A_x - A^0_{\text{In}^-}}$. Determining p$K^\ominus_a$ value through the plot.

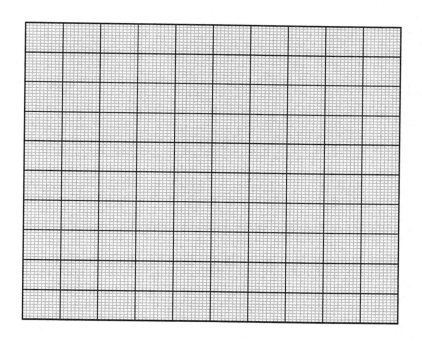

IV. Experimental Error

$$E = \frac{pK_a^\ominus{}_{\text{(experimental)}} - pK_a^\ominus{}_{\text{(theoretical)}}}{pK_a^\ominus{}_{\text{(theoretical)}}} \times 100\% = \underline{\qquad\qquad}.$$

V. Discussion and Exercises

Class_____ Group No. _____ Name_____ Partner_____ Date_____

Experimental Report 7

I. Objects

II. Principles

III. Data Recording and Analyzing

1. Reagents

Reagents	$K_2C_2O_4$	$FeCl_3 \cdot 6H_2O$
Specification		
Mass/g		
Amount of substance/mol		

2. Products

Coordination Compound	Crystal Color	Crystal Appearance	Coordinate Number	Yield / g
$K_3[Fe(C_2O_4)_3]$				

3. Yield Calculation

theoretical product weight (crystal): $W_{theoretical} =$ _____ .

Actual product weight (crystal): $W_{actual} =$ _____ .

Yield: $\dfrac{W_{actual}}{W_{theoretical}} \times 100\% =$ _____ .

IV. Discussion and Exercises

Class_____ Group No._____ Name_____ Partner_____ Date_____

Experimental Report 8

I. Objects

II. Principles

III. Data Recording and Analyzing

1. Determination of Total Hardness of Water

Number	1	2	3
$c(\text{EDTA})/\text{mol}\cdot\text{L}^{-1}$		0.01	
$V_{\text{sample}}/\text{mL}$		50.00	
$V_0(\text{EDTA})/\text{mL}$			
$V_f(\text{EDTA})/\text{mL}$			
$V_1(\text{EDTA})/\text{mL}$	$V_1'=$	$V_1''=$	$V_1'''=$
$V_2(\text{EDTA})/\text{mL}$			
Total hardness/$\text{mg}\cdot\text{L}^{-1}$			

2. Determination of Ca^{2+} Hardness of Water

Number	1	2	3
$c(\text{EDTA})/\text{mol}\cdot\text{L}^{-1}$		0.01	
$V_{\text{sample}}/\text{mL}$		50.00	
$V_0(\text{EDTA})/\text{mL}$			
$V_f(\text{EDTA})/\text{mL}$			
$V_1(\text{EDTA})/\text{mL}$	$V_1'=$	$V_1''=$	$V_1'''=$
$V_2(\text{EDTA})/\text{mL}$			
Ca^{2+} hardness/$\text{mg}\cdot\text{L}^{-1}$			

IV. Error Analyzing

V. Disscussion and Exercises

Class_____ Group No. _____Name_____Partner_____Date_____

Experimental Report 9

I. Objects

II. Principles

III. Data Recording and Analyzing

1. Affection of Temperature

Temperature/°C				
t_0/s (Contol)				
t/s (with 0.2 mg of Mo (VI))				
$\frac{1}{t} - \frac{1}{t_0}$/s^{-1}				

The higher temperature, the faster reaction rate and _____ (longer or shorter) induction time. The induction time of the reaction without Mo (VI) increases _____ (faster or slower) than the reaction with Mo (VI), which means temperature has greater affection to t_0 than activation energy.

2. Working Curve of Mo (VI)

Temperature: _____ °C

Adding Amount of Mo (VI)/μg	0	100	200	400	800
Concentration of Mo (VI)/(μg/mL)					
Induction time t/s					
$\frac{1}{t}$/s^{-1}					

Making the working curve of Mo (VI).

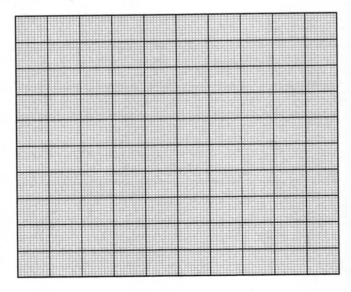

Linear regression equation: _____,
Correlation coefficient $r = $ _____.

3. Determination of Mo (VI) in Synthesized Sample

$t_1 = $ _____ s. $t_2 = $ _____ s.

Concentration of Mo (VI) calculating through the linear regression equation:

Sample	Measured Value (μg/mL)	Average Value (μg/mL)	Relative Error (RE)/%
Synthesized sample			

IV. Error Analyzing

V. Disscussion and Exercises

Class_____ Group No._____ Name_____ Partner_____ Date_____

Experimental Report 10

I. Objects

II. Principles

III. Data Recording and Analyzing

(1) Determination of UV-Vis spectrum of different charged chromium complex ions.

(2) Determine the peak wavelength (λ) of absorbance and its molar extinction coefficient (ε) based on the UV-Vis spectrum of coordinate ions.

Coordinate Ion	$[Cr(H_2O)_4Cl_2]^+$	$[Cr(H_2O)_5Cl]^{2+}$	$[Cr(H_2O)_6]^{3+}$
λ			
ε			

(3) Determination of UV-Vis spectrum of ion exchange elutions of chromium chloride solution.

(4) Determine the relevant contents of coordinate ions in ion exchange elutions of chromium chloride solution by their UV-Vis spectrum.

Coordinate Ion	$[Cr(H_2O)_4Cl_2]^+$	$[Cr(H_2O)_5Cl]^{2+}$	$[Cr(H_2O)_6]^{3+}$
Relevant content			

IV. Error Analyzing

V. Discussion and Exercises